GRAVITATION AND THE UNIVERSE

MEMOIRS OF THE
AMERICAN PHILOSOPHICAL SOCIETY
Held at Philadelphia
For Promoting Useful Knowledge
Volume 78

GRAVITATION
AND THE
UNIVERSE

ROBERT H. DICKE

Cyrus Fogg Brackett Professor of Physics
Princeton University

Jayne Lectures for 1969

AMERICAN PHILOSOPHICAL SOCIETY
INDEPENDENCE SQUARE • PHILADELPHIA

1970

The Jayne Lectures of the American Philosophical Society honor the memory of Henry La Barre Jayne, 1857-1920, a distinguished citizen of Philadelphia and an honored member of the Society. They perpetuate in this respect the aims of the American Society for the Extension of University Teaching, in which Mr. Jayne was deeply interested. When in 1946 this organization was dissolved, having in large measure fulfilled its immediate purposes, its funds were transferred to the American Philosophical Society, which agreed to use them "for the promotion of university teaching, including *inter alia* lectures, publications and research in the fields of science, literature, and the arts."

Accepting this responsibility, the Society initiated in 1961 a series of lectures to be given annually or biennially by outstanding scholars, scientists, and artists, and to be published in book form by the Society. The lectures are presented at various cultural institutions of Philadelphia. Thus far the following, including the series published in the present volume, have been presented:

February 21, 28, March 7, 14, 1961. Per Jacobssen. *The Market Economy in the World of Today*. University Museum, University of Pennsylvania. Memoirs of the American Philosophical Society, Vol. 55 (1961).

March 7, 14, 21, 1962. George Wells Beadle. *Genetics and Modern Biology*. University Museum, University of Pennsylvania. Memoirs of the American Philosophical Society, Vol. 57 (1963).

March 6, 13, 20, 1963. Doris Mary Stenton. *English Justice Between the Norman Conquest and the Great Charter, 1066-1215*. University Museum, University of Pennsylvania. Memoirs of the American Philosophical Society, Vol. 60 (1964).

March 10, 17, 24, 1964. Ellis Kirkham Waterhouse. *Three Decades of British Art: 1710 1770*. Philadelphia Museum of Art. Memoirs of the American Philosophical Society, Vol. 63 (1965).

May 3, 4, 6, 7, 1965. William A. Fowler. *Nuclear Astrophysics*. The Franklin Institute. Memoirs of the American Philosophical Society, Vol. 67 (1967).

February 14, 21, 28, March 7, 1966. Jacob Viner. *The Role of Providence in the Social Order: an Essay in Intellectual History*. University Museum, University of Pennsylvania.

October 31, November 7, 14, 1967. Douglas Bush. *Pagan Myth and Christian Tradition in English Poetry: Three Phases*. Free Library of Philadelphia. Memoirs of the American Philosophical Society, Vol. 72 (1968).

April 4, 9, 11, 1968. Sir Peter Medawar. *Induction and Intuition in Scientific Thought*. University Museum, University of Pennsylvania. Memoirs of the American Philosophical Society, Vol. 75 (1969).

February 13, 18, 20, 1969. Robert H. Dicke. *Gravitation and the Universe*. The Franklin Institute. Memoirs of the American Philosophical Society, Vol. 78 (1970).

Copyright © 1970 by The American Philosophical Society
Library of Congress Catalog Card Number 78-107344

Foreword

Every member of the human race has been aware of some manifestations of gravitation. In our own time, the world has witnessed the incredible translation of Newton's fundamental formulation of the laws of gravitation into a hitherto unimaginable feat—the landing of man on the moon.

More than half a century ago, Albert Einstein conceived and developed a much more far-reaching theory of gravitation, extending the Newtonian concepts to universal applicability. Yet, although the name of this scientific giant is well known in the milieu of the general public, full appreciation of the beauty of Einstein's general theory of relativity is enjoyed by an exceedingly restricted segment of the population. From every point of view, Professor Robert H. Dicke is uniquely qualified for undertaking, in the ninth series of Jayne Lectures, the herculean task of enabling a large group of cultured individuals with widely diversified backgrounds and interests to share in the sense of excitement associated with his own varied and brilliant contributions to this subject—one which represents a truly fundamental approach to the innermost workings of the physical universe.

It is especially appropriate that these lectures should have been held in the halls of the great Institute that stands as a living memorial to that "compleat" scientist, and founder of the American Philosophical Society, Benjamin Franklin. For, Dr. Dicke displays in the contemporary era of science the same evidence of remarkable versatility that characterized Dr. Franklin in an earlier epoch. His multifarious contributions range from the development of radar hardware to the formulation of cosmological theory. He is equally at home with the most pragmatic details of an experiment and with the most esoteric aspects of a conceptual problem. Physicists these days are labelled either as theorists or experimentalists—but Dr. Dicke combines the best of both worlds.

Although special relativity manifests its practical consequences in a number of branches of modern experimental physics, interest in the modern theory of gravitation has been confined largely to the theoretical realms. Shortly after the publication of Einstein's theory in 1916, its apparent success in accounting for certain discrepancies between astronomical observations and the predictions of Newtonian theory seemed to confirm its validity. After a long hiatus, there has been in recent years a resurgence of interest in experimental studies of gravitation, stimulated by recognition of its relationship to other areas of physics and astronomy as well as by the development of new measuring and observing techniques.

In his three Jayne Lectures on "Gravitation and the Universe," Dr. Dicke described in a comprehensible and lucid manner his own recent excursions into these frontiers of intellectual activity. The first dealt with his sophisticated repetition of the Eötvös experiment, in which, improving the accuracy by a factor of a thousand he confirmed to one part in one hundred billion Newton's original assertion about the universality and uniformity of gravitational acceleration—the *principle of equivalence*. Next, he discussed his investigations into the quantitative analysis of the advance of the perihelion of Mercury, prompted by his own work on the modification of general relativity known as the *scalar-tensor theory*. On the basis of his determination of the shape of the sun to an accuracy better than one part in ten thousand, he has called into question the long-standing presumption of perfect agreement between the observations of the behavior of Mercury's orbit and the prediction of Einstein's original expression. Finally, he discussed the cosmological significance of the recently-observed 3°K black-body radiation—the microwave relict of the primeval fireball—predicted by Dr. Dicke and detected by his group at Princeton.

The enthusiastic response of the enthralled capacity audience at each of the 1969 Jayne Lectures at the Franklin Institute attested to Professor Dicke's unassuming and disarming skill in communicating with the students of all ages who enjoyed the privilege of being present.

MARTIN A. POMERANTZ

Author's Preface

When the American Philosophical Society invited me to give the 1969 Jayne Lectures, I accepted with pleasure. A single lecture is often inadequate for the development of some central theme, and I was pleased to be able to give a series of three lectures addressed mainly to a general audience.

The lectures were given in Philadelphia at the Franklin Institute during the evenings of February 13, 18, and 20, 1969, under joint sponsorship by the Institute and the Society. Two central questions tied all three lectures together. First, can one distinguish observationally between General Relativity, Einstein's theory of gravitation, and the scalar-tensor theory, another general relativistic theory closely related to Einstein's theory? Second, how is an experimental physicist to perform significant gravitational experiments when the gravitational interaction is intrinsically so weak that relativistic aspects of the phenomenon are negligibly small in the laboratory? Evidently the first question could not be answered unless the second were first answered.

The great successes of the physicist during the past two centuries have resulted largely from his concentration on one or more narrow goals, and his careful attention to only one aspect of the physical world, an aspect which he has mentally isolated from the rest of the physical world. Thus, by habit and by training, a physicist has limited horizons. His is a laboratory science and in his tiny laboratory he imagines himself cut off from the complexities of the rest of the Universe.

A physicist must put these habits of thought behind him if he is to perform significant gravitational experiments. The earth, the sun, the solar system, even the whole Universe must be his laboratory. But there are grave limitations to experiments performed in such a cosmic laboratory.

Experiments must be carried out with apparatus composed of existing astronomical bodies; deliberate changes in the apparatus are impossible. These large physical systems are extremely complicated and difficult to interpret.

The Physicist's point of view involves an interesting matter of philosophy. Is he permitted to ignore the rest of the Universe in his performance of a laboratory experiment? It is conceivable that ideally the physicist and his apparatus are so inexorably bound to and imbedded in the rest of the Universe that they cannot be separated from it, even conceptually.

Bishop G. Berkeley and E. Mach believed that the only meaningful motion was that of matter relative to other matter. In the development of relativity Einstein was influenced by this and another of Mach's ideas, that the inertial forces observed in an accelerated laboratory are forces induced by distant matter.

In Einstein's theory of gravitation the orientation of inertial coordinate systems relative to the total mass distribution is influenced by this distribution. Except for this somewhat trivial connection, there are no locally observed effects of distant matter of the Universe upon a laboratory experiment under this theory.

The connections between the Universe and the laboratory are somewhat more direct under the scalar-tensor theory of gravitation, but here also non-gravitational physical phenomena are not affected by outside matter. There is one important difference. Under this theory the strength of gravitational interactions steadily decreases relative to other interactions. A scalar field generated by matter everywhere in the Universe, and propagated throughout the Universe, penetrates the walls of the laboratory and influences gravitational experiments through a control of the strength of the interaction. But even here the practical result is minor. The expected rate of change of the gravitational "constant" induced by the slowly changing scalar field is only 1-2 parts in 10^{11} per year. For both theories the rest of the Universe may be practically ignored after the laboratory has been fixed to a coordinate system locally inertial, but from a fundamental point of view the laboratory is inseparable from the Universe.

The sun provides the gravitational field for the experiment described in the first lecture, but the effects of the field are measured in the laboratory. The goal is not one of selecting between the two theories. Rather this precise null experiment provides solid observational underpinning for both theories.

The second lecture was concerned with an old experiment first discussed by Einstein. Again the sun is the static source of gravitational field. Together with the planet Mercury it represents the apparatus. We are concerned with the oblateness of the sun and its significance for the interpretation of the observations. It has long been known that the elliptical orbit of Mercury slowly rotates in space, the effect primarily of gravitational interactions with other planets. Part of the motion has long been interpreted as a relativistic effect, but the observation of a solar oblateness has forced a reconsideration of this interpretation. It is believed that approximately 10 per cent of the motion formerly thought to be relativistic in origin is actually due to the non-relativistic effects of a distorted sun.

The third lecture was concerned with the biggest system of all, the whole Universe. Because of its enormous size and age, relativistic effects are very important. Furthermore, there are several differences in expectation under the two theories. The indications that the Universe developed from an extremely hot fireball are important, for the expectations regarding the formation of helium in the fireball differ significantly for the two theories. A theory of the origin of the globular clusters of stars found in our Galaxy also gives somewhat different results depending upon which gravitational theory is employed.

I wish that I could say that a strong case had been made for or against one or the other of the two gravitational theories, but I cannot. But it was possible to show the manner in which cosmic physics and non-laboratory experiments can be used to investigate this extremely weak interaction. I find it intriguing that a fundamental question of physics can be attacked through diverse and complicated aspects of sciences traditionally regarded as outside the purview of the physicist.

R. H. DICKE

Contents

	PAGE
Foreword by Martin A. Pomerantz	v
Author's Preface	ix
I. The Eötvös Experiment and Gravitation	1
II. Solar Oblateness and Gravitation	26
III. The Cosmic Fireball and Gravitation	49

Illustrations

	PAGE
FIG. 1. A diagram of the type of torsion balance used by Eötvös	6
FIG. 2. The principle of operation of the Princeton version of the Eötvös experiment	8
FIG. 3. Details of the Princeton balance	9
FIG. 4. Details of the vacuum system	9
FIG. 5. The instrument well housing the torsion balance	10
FIG. 6. The instrument seen from the top	11
FIG. 7. A block diagram of the system components	11
FIG. 8. The oscillating-wire detector of small rotations	12
FIG. 9. Tabulated results. The excess acceleration of gold relative to aluminum as a fraction of the net acceleration toward the sun	13
FIG. 10. Distributions of the measurements tabulated in figure 9	14
FIG. 11. The weak equivalence principle	15
FIG. 12. A "gedanken experiment" illustrating the cause of the gravitational red shift	18
FIG. 13. The gravitational deflection of light	27
FIG. 14. The perihelion rotation of the planet Mercury	29
FIG. 15. The solar wind torque	32
FIG. 16. Image of sun, oblateness exaggerated, projected upon the occulting disk	36
FIG. 17. Graphs showing the angle (and time) dependence of the photoelectric current derived respectively from an oblate solar image, and from a displaced solar image	37
FIG. 18. The optical system of the instrument used to measure the solar oblateness	38
FIG. 19. Block diagram of the instrument, showing the main electronic components in relation to the optical system exhibited in figure 18	39
FIG. 20. The upper end of the fixed telescope housing, showing the separately mounted primary mirror	40

Fig. 21. The interior of the compact "instrument shack" in which were housed the telescope and the electronic equipment 41

Fig. 22. The tilt of the sun's axis of rotation 42

Fig. 23. The "diagonal component" of the sun's oblateness, the data of 1966 43

Fig. 24. The same data displayed in figure 23 but broken down into averages over the three different magnifications of the solar image 44

Fig. 25. The 1966 data broken into morning and afternoon groups as a test for a time-dependent systematic error 45

Fig. 26. Globular cluster 53

Fig. 27. Cosmology under modified General Relativity 60

Fig. 28. Similar to curve C of figure 27, except that the gravitational constant is described by the scalar-tensor theory, a general relativistic theory similar to Einstein's theory 63

Fig. 29. The first microwave radiometer, built in early 1945 .. 67

Fig. 30. The original Roll-Wilkinson radiometer housed on top of the Princeton biology building under an old "bird platform." 68

Fig. 31. The three Princeton precision radiometers which operate at the wave lengths 3.2, 1.58, and 0.856 cms. 69

Fig. 32. Spectral distribution expected for thermal (blackbody) radiation at a temperature of 2.7° K. ... 70

Fig. 33. The Princeton isotropy radiometer, constructed by Partridge and Wilkinson 71

Fig. 34. A computed pattern representing the density distribution in the cooling fireball as gas clouds begin to form 73

Fig. 35. The evolution of the hot Universe computed under the scalar-tensor theory 76

Fig. 36. The effect of expansion rate on the production of helium, deuterium, and "helium three" in the cosmic fireball (from Peebles) 78

Fig. 37. Stellar evolutionary ages computed assuming the validity of General Relativity 80

GRAVITATION AND THE UNIVERSE

I. The Eötvös Experiment and Gravitation

INTRODUCTION

Electromagnetism and gravitation are the only known long-range physical interactions, forces exerted by one body on another at a distance. Two classes of short-range forces are also known, the "strong interactions" which, among other things, hold the atomic nucleus together against the disruptive effect of the electrostatic repulsion of its protons, and the "weak interactions" responsible for radioactivity, the spontaneous change of one atomic nucleus into another with the emission of an electron or positron. The "strong interactions" are the strongest known, and the gravitational interaction is the weakest. The gravitational interaction between the proton and the electron in a hydrogen atom is only 4×10^{-40} as strong as the electrostatic force which holds the atom together.

The great weakness of gravitation precludes its experimental investigation by the usual type of laboratory experiment. Traditionally the experimental physicist had been bound to the laboratory. His success has resulted in large measure from concentration on a single localizable aspect of physics and from his isolation in a laboratory shielded from the complexities of the outer world. That this isolation is possible even in principle is the result of a happy combination of three factors: (a) the short-range character of the forces involved, and the resulting localizability of his phenomenon, (b) the possibility, even ease, of eliminating by shielding the disturbances from long-range electromagnetic interactions generated outside the laboratory, and (c) a property of gravitation, an inherently unshielded force, eliminating on the freely falling earth disturbances carried from distant parts of the galaxy by this long-range interaction.

Gravitation is so extremely weak that the cooperation of many atoms in astronomical-sized bodies is required to generate a gravitational field strong enough to permit a useful gravitational experiment. The physicist cannot construct an apparatus on such an enormous scale, certainly not in his laboratory, and he is forced to leave the laboratory and to invade the domain of the astrophysicist and geophysicist if he is to perform a significant gravitational experiment.

Fortunately, our universe contains so many different kinds of objects built on the grand scale that the perceptive experimentalist may be able to find already constructed and working an apparatus capable of teaching him something fundamental about the nature of gravitation. My last lecture will be concerned with the largest of all such apparatuses, the whole Universe itself. We shall see how the structure and history of the Universe are inexorably bound to the nature of gravitation.

The presently interesting gravitational experiments are those involving the principle of relativity. Generally this requires bodies so massive and compact that orbital motion about the body is possible at velocities approaching that of light. Such peculiar objects as pulsars and quasars may be of this type, and we can hopefully expect that in future years significantly new information about the nature of gravitation will result from careful studies of objects such as these. But their great distances introduce formidable obstacles. Most of our information has been obtained through the use of a less suitable but more conveniently located body, our own sun.

The sun is massive but not very compact. A hypothetical planet orbiting close to the sun would have a speed of 420 km/sec or 1/700 the velocity of light. The orbital velocity of the earth is only 10^{-4} times the velocity of light.

Relativistic effects are small in the solar system and great care is required to make measurements of the required precision. Most of the factual information which provides an observational foundation for relativistic theories of gravitation falls into two classes, the group of null experiments and the group of observations which yield a positive result.[1] The "famous three tests" of General Relativity are examples of the latter. It is a mark of Einstein's genius that already in his first paper on General Relativity, his theory of gravitation published in 1916, he was able to suggest all three of these positive tests of his theory.[2] I shall discuss these observations and their modern counterparts in my second lecture.

Tonight I shall be primarily concerned with the Eötvös experiment,[3] one of the null experiments. By a null experiment one means an extremely sensitive test of a theory, an experiment which is expected to give a null result if the theory is correct. In my opinion the corpus of null experiments represents the most significant and important of the observational foundations of gravitational theory.

I do not want to give the impression that Einstein constructed his theory of gravitation as a purely deductive exercise starting from the observations. Such seems to be far from the case. Einstein knew of the few significant observations and he made good use of them. But General Relativity seems to have resulted more from his insight and philosophic outlook than from readily identified specific observations. In his *Autobiographical Notes,* Einstein has attempted to reconstruct, at the age of sixty-seven, the historical setting and perspective that led him to General Relativity.[4] This is a fascinating story which I recommend, but it is not our concern tonight.

Ultimately the justification of a theory must rest on observations. The observations can provide a foundation

---

[1] R. H. Dicke, *The Theoretical Significance of Experimental Relativity* (N. Y., Gordon and Breach, 1964).
[2] A. Einstein, *Ann. d. Phys.* **49**, 769 (1916).
[3] R. v. Eötvös, D. Pekar and E. Feteke, *Ann. d. Phys.* **68**, 11 (1922).
[4] A. Einstein Autobiographical Notes in *Albert Einstein, Philosopher-Scientist,* P. A. Schilpp ed. (N. Y., Tudor Publ., 1949).

for the theory, tests of the theory, but not a proof. A single observation can disprove a theory, but observational proof seems to be impossible.

In considering the Eötvös experiment tonight, the most important of the null experiments, I shall attempt to expose as much of its theoretical significance as possible. I hope to do more than ask the simple question, "Is this experiment in accord with General Relativity?"

In recent years there has been considerable discussion of an alternative general-relativistic theory of gravitation which is also based on Einstein's ideas and has been called the scalar-tensor theory.[5,6] This theory seems to have received a measure of support from serveral observations, including that of the solar oblateness[7] which will be the subject of my second lecture.

In considering the Eötvös experiment and its variations, I shall discuss the experiment, its implications, and I shall ask how much support it provides for both General Relativity[2] and the scalar-tensor theory of gravitation.[16] I shall also describe a crucial variation [8,19] of the experiment still unperformed.

The Eötvös Experiment

There are so many fascinating aspects to this experiment that I scarcely know where to start. First, there is the intriguing chronicle of Roland, Baron Eötvös of Hungary.[9] Almost ensnared by the lure of arctic exploration, a Ph.D. at the age of twenty-two and full professor at twenty-four, Eötvös was one of Hungary's most famous scientists. Second,

---

[5] P. Jordan, *Astr. Nach.* **276** (1948) ; *Schwerkraft und Weltall* (Braunschweig, Wiemeg (1955) ; *Z. Physik* **157** (1959) : p. 112.

[6] C. Brans and R. H. Dicke, *Phys. Rev.* **124** (1961) : p. 925; R. H. Dicke, *Phys. Rev.* **125** (1962) : p. 2163.

[7] R. H. Dicke and H. M. Goldenberg, *Phys. Rev. Letters* **18** (1967) : p. 313.

[8] R. H. Dicke in *Gravitation and Relativity*, H. Y. Chiu and W. F. Hoffmann eds. (N. Y., Benjamin, 1964).

[9] R. v. Eötvös, *Gesammelte Arbeiten*, P. Selenyi ed. (Budapest, Akad. Kaido, 1953).

there is the history of this old experiment. The roots of the Eötvös experiment probably go back to ancient times and the list of early experimenters includes the names of such distinguished scientists as Galileo, Newton, and Bessel.[10] Third, there is the delicacy of the modern experiment and its implication. Few experiments are simpler in principle, harder in practice, and so far-reaching in implication.

The Eötvös experiment is concerned with the following question: Does the gravitational acceleration of a body depend upon its composition?

It seems likely that one of our ancient forebears, slightly more perceptive than most, might have performed the first crude version of this experiment when in knocking his breakfast from a tree with a stone he happened to note that both the squirrel and stone hit the ground at about the same time.

Galileo for his version may not have actually performed his famous experiment, simultaneously dropping wood and iron balls from the leaning tower of Pisa. But it hardly matters, for his other experiments using rolling spheres were equivalent and he was the first to state clearly the conclusion.

Sir Isaac Newton was a great theorist, but his version of this experiment is elegant, and it is so simple that it can be performed easily at home using nothing more complicated than empty cans and wire. The recipe is simple: Suspend two cans by long pieces of wire so that each swings as a pendulum with the open end up. Fill both cans to the top with water and set them both swinging, having first adjusted their periods to be approximately equal. Time the intervals between phase coincidence. Now empty one can and refill it with oil, sand, or some other material and repeat the timing procedure. With an accuracy of a few tenths of a per cent, it will be found that the period is independent of the material.

---

[10] F. W. Bessel, *Pogg. Ann.* 25 (1832) : p. 401.

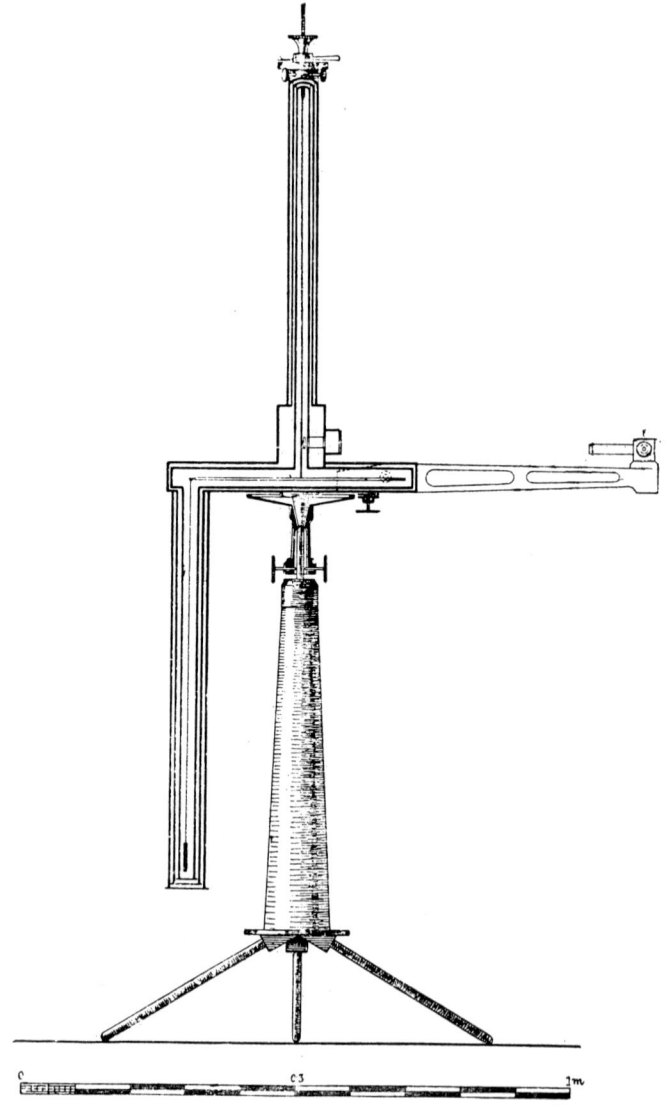

Fig. 1. A diagram of the type of torsion balance used by Eötvös.

Eötvös' 1909 version of the experiment was 10^5 times as accurate as Newton's.[3] Employing a static torsion balance with different materials suspended from the ends, Eötvös, Pekar, and Feteke improved the accuracy to 5 parts in a billion. Figure 1 shows an old illustration of one of Eötvös' torsion balances.

The materials examined by Eötvös include such exotic substances as asbestos, talc, and snakewood, but we now recognize that elements differing substantially in elementary particle content, nuclear binding energy, electrostatic energy, and particle velocity are to be intercompared if fundamental questions are to be answered. Heavy and light elements are most suitable. Fortunately, Eötvös included both platinum and light elements among his samples.

The only modern version of the Eötvös experiment was performed by Roll, Krotkov, and me.[11] Eötvös had neither high-vacuum techniques nor electronic instruments for his experiment. I had thought that the best of modern techniques should be capable of an easy two orders of magnitude improvement. We did succeed in achieving an increase in precision by a factor of 500 but it was not easy and a long time was required.

Eötvös employed the earth as the source of his gravitational field whereas we used the sun. Under the gravitational pull of the sun, the earth, laboratory, and sample weights all accelerate toward the sun at 0.6 cm/sec². We were able to show with an accuracy of a part in 10^{11} that the accelerations of gold and aluminum are equal. The two accelerations are not likely to differ much more than 6×10^{-12} cm/sec². It can be reported that a body accelerating from rest at this rate for 3000 years would reach the magnificent velocity of 6 mm/sec.

The principle of the experiment is shown in figure 2. The weights are so positioned at 6 A.M. that any tendency

---

[11] P. G. Roll, R. Krotkov and R. H. Dicke, *Ann. of Physics* **26** (1964): p. 442.

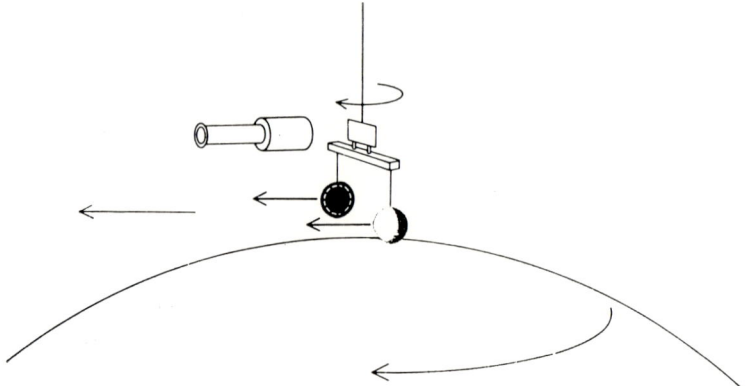

FIG. 2. The principle of operation of the Princeton version of the Eötvös experiment. At 6 o'clock both test spheres accelerate horizontally toward the sun. If the acceleration is greater for one sphere, the balance will rotate.

of the weights to fall unequally toward the sun would cause a rotation of the torsion balance. Twelve hours later the rotation would be in the opposite direction. The torque required to keep the balance from rotating is recorded and analyzed. Any 24-hour period in the torque can thus be detected.

The form of balance actually employed is shown in figure 3. To minimize disturbances due to a shifting mass distribution on the earth's surface, the balance employs three weights instead of two, one gold and two aluminum.

To reduce extraneous, disturbing torques to sufficiently low values, the balance was placed in a highly evacuated chamber in a thermally stable, deep instrument well far removed from a building. Even with a good vacuum, thermal constancy and uniformity are required. It is easily calculated that in the vacuum chamber temperature difference between opposite sides greater than 10^{-5} °K may be excessive.[11] See figures 4-6.

Figure 7 shows schematically the main elements of the apparatus. In the upper left corner is shown the triangular torsion balance seen looking down. The gold weight is

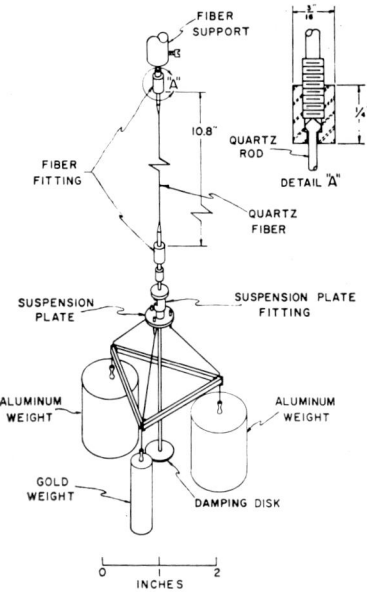

FIG. 3. Details of the Princeton balance. Three weights are used to impart a 3-fold symmetry axis to the balance.

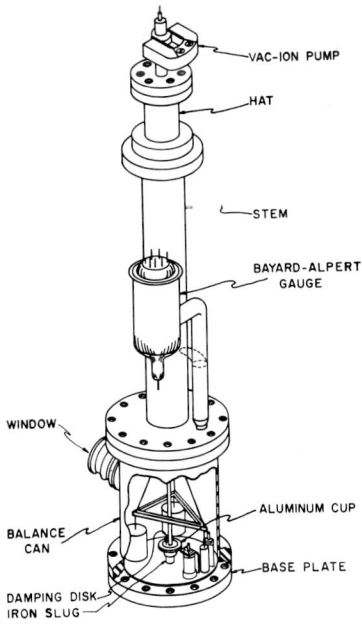

FIG. 4. Details of the vacuum system.

Fig. 5. The instrument well housing the torsion balance. In operation the well is un-manned and is capped with a 3-foot insulated plug carrying embedded electric blankets.

acted on by an extremely weak, electrically induced torque. The sign and strength of this torque are derived from the angular position of the balance through the feedback circuit. This torque opposes other externally produced torques applied to the torsion balance to prevent automatically a rotation of the balance. By including "velocity

THE EÖTVÖS EXPERIMENT AND GRAVITATION 11

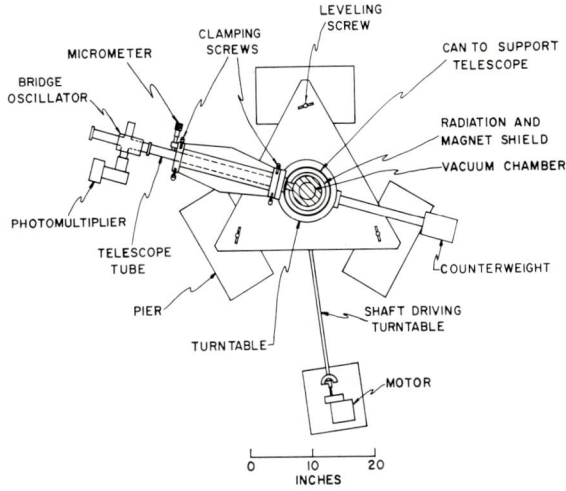

FIG. 6. The instrument seen from the top.

feed-back" the free oscillation of the balance is strongly damped. This has the very useful effect of quickly eliminating any externally generated disturbance of the balance, as by a passing truck, causing the oscillation to be damped in a few minutes rather than in many hours.

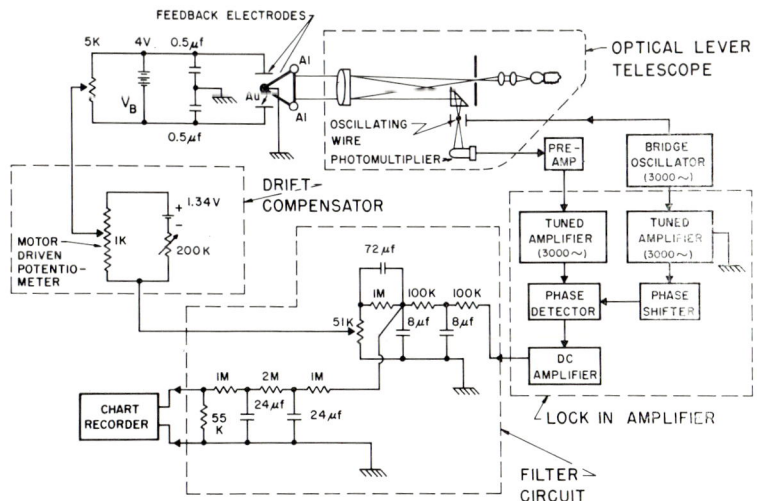

FIG. 7. A block diagram of the system components.

The details of the rotation detection system are shown in the upper right of figure 7. Light from an illuminated slit is reflected from the polished face of the quartz triangle. This reflected light is focused by the achromatic objective lens to form an image on an oscillating wire of the rotation detector. See figure 8. Light passing the oscillating wire is detected by a photomultiplier.

When the slit image is accurately centered on the oscillating wire, the light flux streaming by the wire is modulated at twice the oscillation frequency of the wire. If the image is not accurately centered, the fundamental frequency also appears in the modulation. This fundamental component is detected and provides a measure of the rotation of the torsion balance. This technique is extremely sensitive and a detection of a rotation as small as 10^{-9} radian can be achieved.

On the left side of figure 7 is shown an element called the drift compensator. Even after months of operation the fine quartz fiber of the torsion balance continues slowly

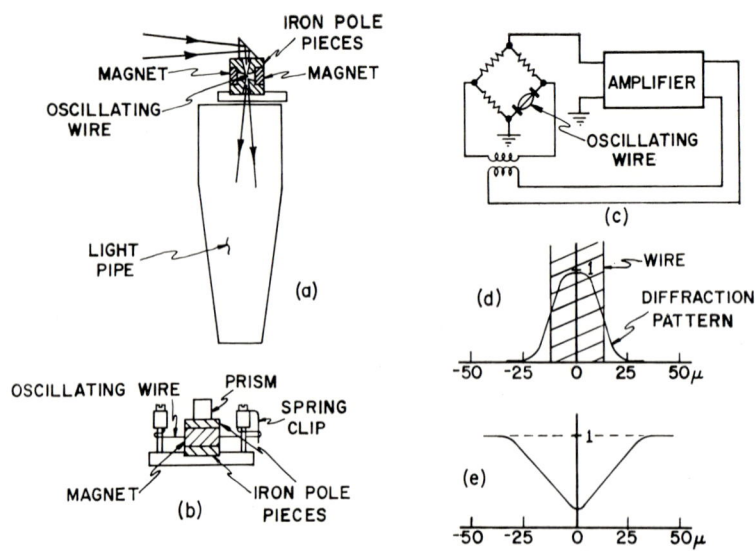

Fig. 8. The oscillating-wire detector of small rotations.

THE EÖTVÖS EXPERIMENT AND GRAVITATION 13

MEAN VALUES OF η(Au, Al), WITH PROBABLE ERRORS OF THE MEAN

Points more than 3 std. dev. from mean excluded	Telescope orientation	No temperature regression subtracted ($B(T2) = 0$)	T2 regression subtracted ($B(T2) = 6.3 \times 10^{-8}$ stu/°C)	T5 regression subtracted ($B(T5) = 5.4 \times 10^{-8}$ stu/°C)
No	North	$(+2.22 \pm 1.42) \times 10^{-11}$	$(1.38 \pm 1.27) \times 10^{-11}$	$(1.30 \pm 1.25) \times 10^{-11}$
No	South	$(-2.53 \pm 1.94) \times 10^{-11}$	$(-0.25 \pm 2.70) \times 10^{-11}$	$(1.33 \pm 1.73) \times 10^{-11}$
No	North + South	$(-0.09 \pm 1.20) \times 10^{-11}$	$(0.59 \pm 1.45) \times 10^{-11}$	$(1.32 \pm 1.04) \times 10^{-11}$
Yes	North	$(+2.22 \pm 1.42) \times 10^{-11}$	$(1.38 \pm 1.27) \times 10^{-11}$	No values more than 3 std. dev. from mean
Yes	South	$(-0.59 \pm 1.51) \times 10^{-11}$	$(2.44 \pm 2.00) \times 10^{-11}$	
Yes	North + South	$(+0.89 \pm 1.03) \times 10^{-11}$	$(1.94 \pm 1.15) \times 10^{-11}$	

FIG. 9. Tabulated results. The excess acceleration of gold relative to aluminum as a fraction of the net acceleration toward the sun.

and steadily to untwist. This steadily increasing minute torque is compensated by a programmed electrically induced torque.

The instrument was sealed in the instrument well and operated remotely from a distant instrument shack for months at a time (see figure 5). As remarked above, we found with an accuracy of 10^{-11} that the accelerations toward the sun of aluminum and gold are equal. The results of measurements made over a nine-month period are tabulated in figure 9. The first column summarizes the results ignoring the effect on the balance of a twenty-four-hour period in the temperature of the instrument well. The numbers tabulated represent the fractional excess of gold over aluminum in the gravitational accelerations toward the sun. The second and third columns represent the results after subtracting the effect of temperature variation as measured by two differently located thermisters.

The distributions of individual measurements are shown on the right side of figure 10. The first two columns refer to sine and cosine components of the 24-hour period in the feed-back voltage applied to the balance. The last column represents the distributions of individual measurements for which average values are tabulated in the first two rows of figure 9.

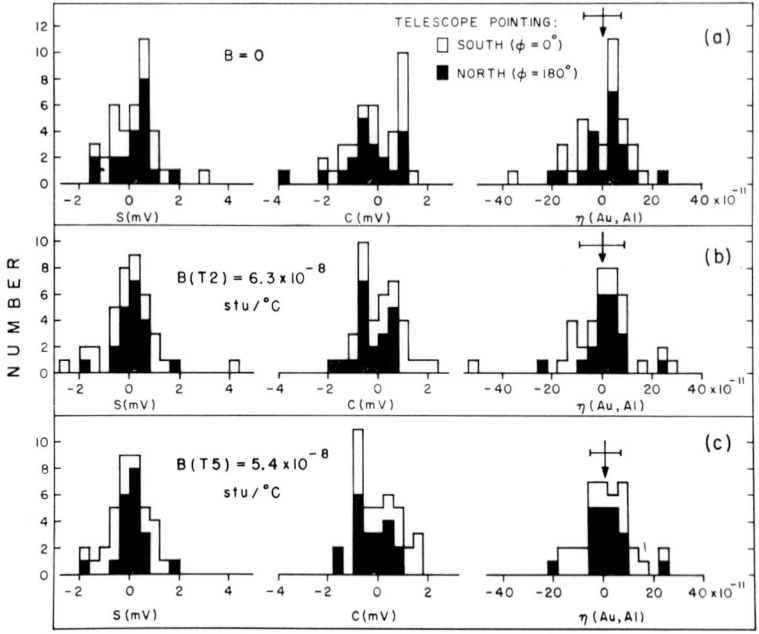

Fig. 10. Distributions of the measurements tabulated in figure 9. (The distributions at the right.)

The Significance of the Eötvös Experiments

In Einstein's theory, gravitation is described by a single field quantity, a tensor, which is usually but not always interpreted as the metric tensor of the geometry of spacetime. Under this theory, sufficiently small test bodies fall with a unique acceleration. In fact this result is central to the whole theory, being a manifestation of the Equivalence Principle, a pillar of the theory.

To be more precise, the constancy of the gravitational acceleration is an expression of the "weak equivalence principle," whereas Einstein's gravitational theory is rooted in the "strong principle."[1, 8, 12] By the "weak equivalence principle" one means the equivalence over a small region in space of gravitational and accelerational effects. In a freely falling elevator, with the elevator and all of its con-

---

[12] R. H. Dicke, *Science* **129** (1959): p. 621.

tent falling with the same acceleration, the gravitational force has disappeared. This idea is illustrated in figure 11, meant to represent the generation of a gravity-like force in a small laboratory through the simple expedient of accelerating the laboratory with a spring.

By the "strong equivalence principle" one means that in a freely falling and non-rotating laboratory the laws of physics, including their numerical content, are the same everywhere including gravity free space. Thus the "strong principle" says more than the "weak principle," but it is only the "weak principle" that is directly supported by the Eötvös experiment.

Einstein's General Relativity is so constructed that it implies that in a small region of space-time in an appropriate coordinate system, i.e. inside a freely falling, non-rotating laboratory, the physical laws take on a

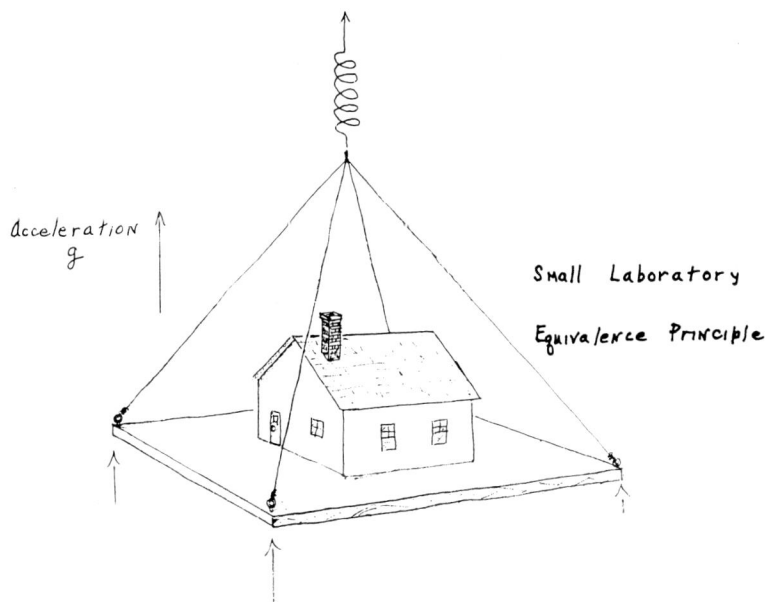

Fig. 11. The weak equivalence principle. Physical phenomena in a uniformly accelerated laboratory in gravity-free space are similar to those experienced in a gravitational field.

standard form and numerical content. Thus "strong equivalence" is satisfied. The description of gravitation by a single symmetrical metric tensor provides the reason for "strong equivalence." The "appropriate coordinate system" is one which induces a standard form in the metric tensor at the location of the elevator, the standard form being that of an inertial coordinate system in gravity free space.

The scalar-tensor theory differs from General Relativity by the inclusion of a scalar field in addition to the tensor. This scalar has the effect of locally controlling the strength of the gravitational interaction. Thus the "strong equivalence principle" is not satisfied, for the "gravitational constant' is not constant in this theory and, expressed in dimensionless form, this "constant" is part of the numerical content of physical laws.

The scalar field in this theory provides a possible relativistic mechanism for affecting more than the gravitational constant. Thus, for example, the mass ratio for the proton and neutron or the charge of the electron might be functions of the scalar.

It is interesting that the Eötvös experiment, which is directly concerned only with the "weak equivalence principle," nonetheless also gives us useful information about the "strong principle." Consider the following example: Assume for the moment that the mass of the neutron increases with height (relative to the proton and electron) and consequently that the internal energy of an atom containing neutrons increases relative to a neutronless atom (hydrogen). This implies that, per unit inertial mass, more work would be required to lift such a neutron-rich material than would be required to lift hydrogen. The neutron-rich material should have an anomalously large weight for its mass, i.e. it should fall with an anomalously large gravitational acceleration. The observation of Eötvös, that light and heavy atoms fall with very nearly the same gravitational acceleration, sets extremely stringent limits to possible variability with height (or with a scalar field) of the neutron to proton mass ratio.[13, 14] If any such variation were to exist, it must be negligibly small.

This type of argument can be extended, and with the improved precision of the Princeton experiment I conclude that only the gravitational "constant" and the weak interaction "constant" could be significantly dependent upon a scalar field. Among the effects which can be excluded are variations in the electronic charge and the strong interaction constants.

The above argument and conclusion is so important that I shall develop it with a bit more care. The argument is illuminated by a "gedanken experiment" based on three key assumptions:

(1) The inertial mass of a body is proportional to its total energy content, the famous relation $E = Mc^2$. This is a central part of Einstein's special relativity theory which receives massive support from many laboratory experiments, particularly those involving elementary particles at high energies.

(2) Heavy and light atoms fall with the same gravitational acceleration (the Eötvös experiment).

(3) Energy is conserved for a closed system in a static gravitational field.

The "gedanken experiment" requires "gedanken" apparatus shown in figure 12.

Imagine the following sequence of events: (1) A photon ejected from an energy reservoir excites an atom from its ground-energy state to an excited state. (2) An energy source in the reservoir is then used to lift the excited atom. (3) A photon is emitted by the excited atom and absorbed in the reservoir. The atom is thus returned to its ground energy state. (4) Finally, the atom is lowered to its initial position.

---
[13] A. H. Wapstra and G. J. Nijgh, *Physica* 21 (1955) : p. 796.
[14] R. H. Dicke in *Rendiconti della Scuola Internazionale di Fisica "Enrico Fermi,"* XX Corso. C. Møller, ed. (N. Y., Academic Press, 1961).

GRAVITATIONAL RED SHIFT

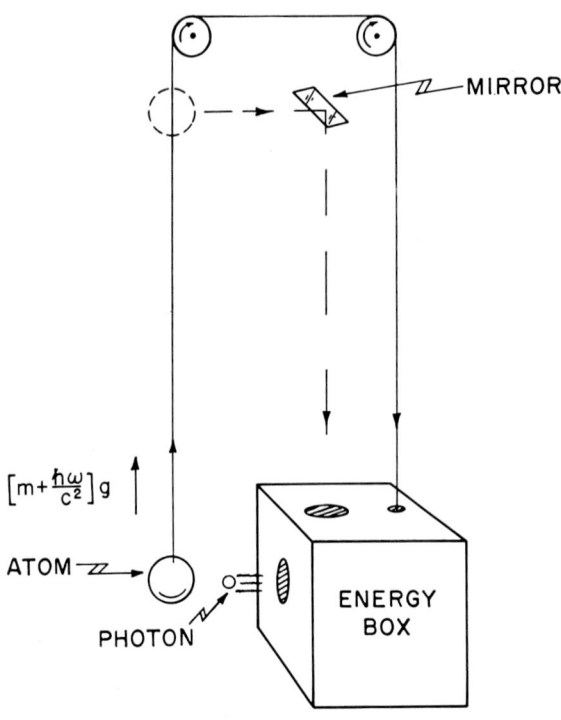

Fig. 12. A "gedanken experiment" illustrating the cause of the gravitational red shift. The same type of "gedanken" apparatus is used to discuss the relations between the Eötvös experiment and the strong equivalence principle.

The atom is left in its original ground-energy state and all energy is obtained from and returned to the reservoir. The net transfer of energy from the reservoir must be zero if energy is to be conserved. But the initial excitation of the atom by the photon increases its energy, hence its mass and weight. Thus more work is required to lift the atom in its excited state than is returned to the energy reservoir when the atom is lowered in its ground state. For energy to be conserved, the photon returned to the box must be more energetic (bluer) than the photon taken from the box. This increase in energy of the returned photon is equal to the energy required to lift a mass corresponding to the photon energy.

The blue shift, or red shift of a photon emitted at the bottom relative to the one emitted at the top, was an early prediction of Einstein. Its observation in solar spectral lines has had a long but ambiguous history. Only in recent years have technical difficulties been overcome to show the expected red shift with an accuracy of 5 per cent.[15] The red shift has also been demonstrated, with greater precision, in the laboratory using Mössbauer gamma rays.[16]

How is this "gedanken experiment" relevant to the strong equivalence principle?[1] The constancy of the numerical content of physical law implies that the excitation energy of an atom is constant. Imagine that the converse were true, that the excitation energy of the atom were to increase with height, implying that the photon returned to the box would be even more energetic. For energy to be conserved this increased energy of the photon emitted at the top must be reflected by an increase in the work required to lift the excited atom. Assume that the weight F of the atom in its ground state is given by its mass M_o

$$F_o = M_o g = E_o g/c^2, \qquad (1)$$

where E_o refers to the inertial mass and total energy of the atom respectively; g and c are the acceleration of gravity and the velocity of light. The weight of the excited atom is

$$F = E_1 g/c^2 + \frac{dE_1}{dh} = (E_o + \Delta E)g/c^2 + d(\Delta E)/dh, \qquad (2)$$

where E_1 represents the total energy of the atom in its excited state. $\Delta E = E_1 - E_o$ is the excitation energy of the atom and the last term of equation (2) represents the rate of change of this energy with height h. This last term is the

---
[15] J. Brault, Dissertation, Dept. of Physics, Princeton University (1962).
[16] R. V. Pound and G. A. Rebka, *Phys. Rev. Letters* 4 (1960) : p. 337; R. V. Pound and R. L. Snider, *Phys. Rev.* 140 (1965) : p. B788.

anomalous weight associated with the change in excitation energy with height. This anomalous weight causes an anomalous gravitational acceleration[1, 14]

$$\delta g = (c^2/E_1)(dE_1/dh). \quad (3)$$

Whatever the cause, a change of the total internal energy of an atom with height introduces an anomalous weight, and the anomalous acceleration is given by eq. (3). Assuming that the Princeton version of the Eötvös experiment requires $|\delta g/g| < 2 \cdot 10^{-11}$, we conclude that

$$\left| \frac{dE_1}{dh} \right| < 2 \times 10^{-11} E_1 g/c^2. \quad (4)$$

This type of argument can be extended to cover a variety of situations. Assume for the moment that the square of the elementary electronic charge, e^2, is a function of the gravitational scalar of the scalar-tensor gravitational theory. (Actually, this way of stating the assumption is careless and inexact. The elementary charge is a physical constant represented by a number carrying physical dimensions. It can be varied arbitrarily by changing the units of measure provided by the United States Bureau of Standards. Only dimensionless numbers are free of this ambiguity. The fine structure constant $e^2/\hbar c \cong 1/137$ is such a number. If Planck's constant $\hbar$ and the velocity of light c are defined to be constant, i.e. the units of measure are so chosen as to keep them constant, the constancy of the dimensionless fine structure constant is equivalent to the constancy of the elementary electronic charge. It is in this sense that I speak of charge being constant.) The energy stored in the electric field about the atomic nucleus is proportional to the square of the nuclear charge. Thus this contribution to the total energy of the atom is disproportionately large for an atom with a large nuclear charge. As the discussion below shows, the implication drawn from the Princeton version of the Eötvös experiment is that very little variation of the elementary electronic charge with height would be possible, and for a reasonable field theoretic treatment of the question, significant variations with either space or time can be excluded.

The scalar field ϕ of the scalar-tensor theory of gravitation[6] is a function of distance from a massive body, being increased fractionally through the presence of the body by the amount

$$\delta\phi/\phi = GM / (\omega + 2) Rc^2, \tag{5}$$

where M and R are the mass and distance to the body. ω is a dimensionless constant approximately equal to 5.

For a gold atom the electrostatic energy represents 4×10^{-3} of its total energy whereas for aluminum it represents only 4×10^{-4} of the total. Thus the anomalous gravitional acceleration would be greater for gold. From eq. (3) the difference in the two accelerations would be ($\delta g = g_{Au} - g_{Al}$)

$$\delta g = 3.6 \times 10^{-3} \frac{c^2}{e^2} \frac{de^2}{dh} . \tag{6}$$

From eq. (4)

$$\frac{1}{e^2} \left| \frac{de^2}{dh} \right| < \frac{2.10^{-11}}{3.6 \cdot 10^{-3}} \frac{g}{c^2} \tag{7}$$

$$\frac{1}{e^2} \left| \phi \frac{de^2}{d\phi} \right| \cdot \left| \frac{1}{\phi} \frac{d\phi}{dh} \right| < 5.5 \cdot 10^{-9} \frac{g}{c^2} . \tag{8}$$

From eq. (5)

$$\frac{1}{\phi} \frac{d\phi}{dR} = - \frac{GM}{(\omega+2) R^2 c^2} = - \frac{g}{(\omega+2) c^2} ,$$

eqs. (8) and (9) give (for $\omega = 5$)

$$\left| \frac{\phi}{e^2} \frac{de^2}{d\phi} \right| < 5.5 \cdot 10^{-9} (\omega + 2) \sim 4 \times 10^{-8}. \tag{10}$$

From eq. (10), the elementary electronic charge cannot be a sensitive function of the scalar of the scalar-tensor theory if the above assumptions are satisfied. The limit imposed by eq. (10) is a million times smaller than the type of variation which might reasonably be expected if there should be a coupling of the electronic charge to a scalar field. We conclude that the elementary electric charge is very likely independent of a scalar gravitational field.

Equation (10) can also be used to set stringent limits to a possible variation of the elementary electrical charge e with time. Under the scalar-tensor theory of gravitation the scalar increases slowly with time.[6] For a cosmology characterized by a cosmologically flat space, a universe which expands without limit but steadily and continuously slows down in its expansion rate, the scalar increases fractionally at the rate

$$\frac{1}{\phi}\frac{d\phi}{dt} = \frac{2}{(3\omega+4)T}, \qquad (11)$$

where T represents the age of the Universe. (It is generally believed that the Universe does not differ drastically from the flat-space form.) Multiplying eqs. (10) and (11) yields

$$\left|\frac{1}{e^2}\frac{de^2}{dt}\right| < \frac{1.1 \times 10^{-8}}{T}\left(\frac{\omega+2}{3\omega+4}\right) \qquad (12)$$

$$< \sim 10^{-19}/\text{year},$$

assuming that $T \sim 7 \times 10^9$ years. This rate of change is negligibly small.

The argument based on the Eötvös experiment, for the constancy of the electronic charge carries more precision than more direct arguments such as those based on radioactive decay in ancient times.[17]

I conclude from the above arguments based on null result of the Eötvös experiment and the well-established principles of mass-energy equivalence and energy conservation that little if any variation with position of the elementary electronic charge can occur. Such a variation, if it occurred,

---

[17] P. J. E. Peebles and R. H. Dicke, *Phys. Rev.* **128** (1962): p. 2006.

should be associated with coupling to a scalar, such as a gravitational scalar, field. But the relation between time variation and spatial variation of such a field imposes a limit on a possible variation with time of electronic charge. Both variations are negligibly small. It is unlikely that any significant variation with either time or space of the electronic charge should occur.

Similar arguments can be used to eliminate variations of several other dimensionless physical constants such as the ratios of the masses of elementary particles and the strong coupling constant responsible for the binding of the atomic nucleus. Thus, much of the strong equivalence principle is supported by the Eötvös experiment. However, two important parts of this principle are left unsupported, the constancy of the weak interaction constant and the constancy of the gravitational coupling constant. This latter question will be our final concern this evening.

For an ordinary laboratory-sized spherical weight the ratio of gravitational energy to total energy is

$$\frac{3}{5}\left(\frac{GM}{Rc^2}\right) \sim 10^{-27}, \qquad (13)$$

completely negligible. Hence, the question of variability of the gravitational coupling constant with position cannot be investigated with an ordinary laboratory-type Eötvös experiment. Such a variability is the principal prediction of the scalar-tensor theory.

What about an Eötvös experiment on an astronomical scale? It has been suggested[18] that an anomaly in Jupiter's motion associated with a variation of its gravitational self-energy might someday be measurable, and Nordtvedt[19] proposed using the Trojan asteroids as comparison test bodies to observe Jupiter's anomalous acceleration.

---

[18] See reference cited in note 8: append. A, chap. 1.
[19] K. Nordtvedt, Jr., *Phys. Rev.* **169** (1968): p. 1014; *Phys. Rev.* **170** (1968): p. 1186.

In similar fashion an anomalous acceleration of the earth toward the sun should cause a small effect on the twenty-four-hour period in earth tides measured by sensitive gravimeters. The small weight actuating the gravimeter would fall to the sun without this anomalous acceleration. This effect was investigated a half-decade ago with the conclusion that the small correction to the twenty-four-hour period would be lost in other unpredictable effects. Nordtvedt[19] has suggested the use of the moon as a comparison test body. Under the scalar-tensor theory the moon's orbit should be slightly shifted along the earth-sun axis. The shift is small, but it may be measurable through the use of the technique of optical-ranging.[20] Cube-corner reflectors could provide a fixed reference point on the moon's surface and pulsed laser beams could be used to measure the distance.[21]

Under the scalar-tensor theory the anomaly of the gravitational acceleration due to the variation of gravitational self-energy is given by eq. (3) with the subscript on E dropped. Here E refers to the total self-energy of the test body. Combing eqs. (3) and (9) gives [1, 14]

$$\delta g = \frac{c^2}{E} \left(\phi \, \frac{dE}{d\phi} \right) \left(\frac{1}{\phi} \, \frac{d\phi}{dR} \right) = -\frac{g}{(\omega + 2)} \left(\frac{\phi}{E} \right) \left(\frac{dE}{d\phi} \right). \tag{14}$$

Nordtvedt[22] has approached this problem in a different way, through the equations of motion.

Under the scalar-tensor theory, the gravitational constant varies inversely with the scalar ϕ. Hence, for an astronomical body negligibly compressed by gravity the fractional anomaly in its gravitational acceleration is

$$\frac{\delta g}{g} = + \frac{1}{\omega + 2} \cdot (E_g/E). \tag{15}$$

---

[20] W. F. Hoffmann, R. Krotkov and R. H. Dicke, *I. R. E. Trans. on Mil. Electr.* **MIL-4** (1960) : p. 28; R. H. Dicke, W. F. Hoffmann and R. Krotkov in *Space Research II*, eds. H. C. van de Hulst, C. de Jager and A. F. Moore (Amsterdam, North-Holland Publ. Co., 1961).

[21] C. O. Alley, P. L. Bender, R. H. Dicke, J. E. Faller, P. A. Franken, H. H. Plotkin and D. T. Wilkinson, *Jour. of Geophys. Res.* **70** (1965) : p. 2267.

[22] K. Nordtvedt, *Phys. Rev.* **169** (1968) : p. 1017.

Here E_g is the body's (negative) gravitational energy. For the earth the fractional variation of the radius is approximately one-tenth of the fractional variation of ϕ. Thus,

$$\frac{\delta g}{g} \cong \frac{0.9}{\omega + 2} \cdot (E_g/E). \tag{16}$$

For an object like the sun, supported internally by hot gas, the total (positive) binding energy is one-half of the negative of the gravitational energy. For this case

$$\frac{\delta g}{g} = \frac{0.5}{\omega + 2} \cdot (E_g/E). \tag{17}$$

For all cases the anomaly is negative, i.e. the gravitational acceleration is reduced.

What is the central conclusion? A tale having its roots in ancient times but extending to the present should contain a maxim, but I have not found it! Perhaps it is that a remarkably simple experiment carried out with enough precision is capable of yielding important results. Perhaps it is that a profound conclusion can be hidden inside a simple observation. Perhaps it is that a physicist should be bold; if an ordinary laboratory apparatus is too small to detect a weak effect, he should abandon the laboratory and take the earth, solar system, galaxy, or Universe to be his apparatus.

It has been noted that gravitation, the weakest of all known interactions, requires non-laboratory apparatus. The Princeton version of the Eötvös experiment, employing the sun as a source of gravitational field, supports the weak principle of equivalence and most of the strong principle. But it is just the unsupported part of the strong principle that represents the primary distinction between General Relativity and the scalar-tensor theory of gravitation. A yet-to-be performed version of the Eötvös experiment, employing astronomical-sized bodies as test particles, is capable of exhibiting a difference, but this might provide a subject for a future talk.

II. Solar Oblateness and Gravitation

As I emphasized in my last lecture, gravitation is an extremely weak interaction and significant experimentation requires astronomical bodies as sources of the field. We discussed the Eötvös experiment, one of those highly precise null experiments which have provided the most important part of the observational basis for gravitational theory. Tonight I consider a problem associated with another class of observations, the so-called "positive tests of General Relativity." These observations are very few in number, relatively inaccurate, but particularly significant. They provide means for distinguishing observationally between General Relativity and the scalar-tensor theory of gravitation.

Perhaps the best known of the "positive tests of General Relativity," the *gravitational red-shift,* was discussed last week. While this experiment provides a positive result, the experiment is not, strictly-speaking, a "positive test" specifically for General Relativity. The same result is expected under the scalar-tensor theory of gravitation and is derived from simple considerations not requiring the machinery of a complete gravitational theory.

Only two observations are presently known which provide specific tests for General Relativity. Both were discussed in Einstein's[1] comprehensive paper on General Relativity, and the first of these, the gravitational deflection of light, was first observed in 1919[2] three years after the publication of that paper.

The classical method of observing the gravitational deflection of light requires a total eclipse of the sun. The star field seen about the eclipsed sun is photographed during the few moments the eclipse is total, and star positions

---
[1] A. Einstein, *Ann. d. Phys.* **49** (1916): p. 769.
[2] F. W. Dyson, A. S. Eddington, and C. Davidson, *Phil. Trans. Roy. Soc.* (London) **220A** (1920): p. 291.

are accurately measured on the developed plates. Owing to the short duration of the eclipse and the consequent absence of repetitions of the observation, there has always been considerable doubt about the freedom of the final results from systematic errors. Furthermore, the results derived from past solar eclipses (figure 13) have scattered a great deal. The accuracy of the gravitational deflection of light determined from total eclipses is probably no better than 20 per cent.

Recently several new techniques have appeared that seem destined greatly to improve the situation. The first of these is a technique devised by I. Shapiro.[3] It makes use of the wave retardation associated with the deflection of a wave indirectly to detect the presence of the deflection. This observation is carried out by radar, waves being sent to the planets Venus and Mercury when they are near the back side of the sun. The delay suffered by the wave in passing close to the sun is observed. Only preliminary results are

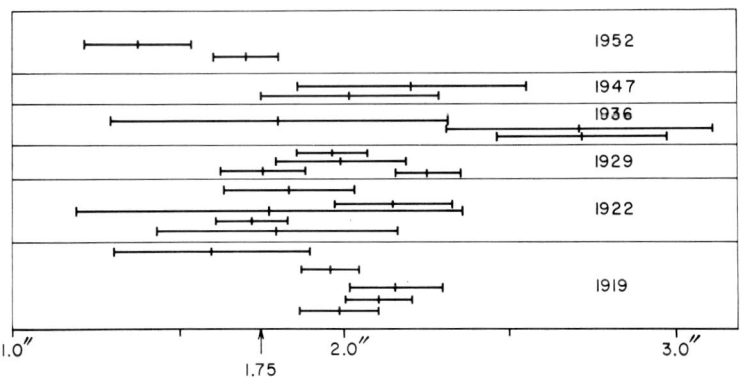

FIG. 13. The gravitational deflection of light. Einstein predicted a 1.75″ arc deflection for a light ray passing close to the sun's limb. All these results are based on studies of photographs of total eclipses of the sun.

---

[3] I. I. Shapiro, G. H. Pettingill, M. E. Ash, M. L. Stone, W. B. Smith, R. P. Ingalls, and R. A. Brockelman, *Phys. Rev. Letters* **20** (1968) : p. 1265.

available thus far, but they are already substantially better than the classical observations.

Another approach which seems promising makes use of a new technique devised by H. Hill. This employs a new specialized instrument capable of observing stars in the sky close to the sun even without the benefit of an eclipse. This apparatus is not yet completed, but hopefully it should give observations on stars on such a regular and continuing basis that the difficulties associated with systematic errors can be investigated and eliminated.

Still another possible approach makes use of long-baseline radio interferometry to measure the position of small radio sources when near the sun.[3] Strangely enough, measurements made at wave lengths 10^4-10^5 times as long as optical radiation are capable of higher resolution and better angular measures than those based on light rays.

Under the scalar-tensor theory of gravitation the deflection expected is[4] $(1-s) \times$ (Einstein's value), where $s = 1/(2\omega + 4)$ represents the fraction of a body's weight due to the scalar interaction.[5] For $\omega = 5$, a value consistent with the solar oblateness observations to be discussed, the expected deflection would be 93 per cent of Einstein's value.

The gravitational deflection of light is an important relativistic effect capable of distinguishing between ordinary general relativity and the scalar-tensor theory, but this is not our primary concern today. The only other "positive test" observed thus far is not properly a prediction of Einstein although he was the first to give a believable explanation for it. From the early nineteenth century a discrepancy in the motion of the planet Mercury had been known, an effect that could not be accounted for using ordinary Newtonian gravitational theory. The effect in question, an excess rotation of the perihelion of the orbit

[4] C. Brans and R. H. Dicke, *Phys. Rev.* **124** (1961) : p. 925.
[5] R. H. Dicke, *Phys. Rev.* **125** (1962) : p. 2163.

SOLAR OBLATENESS AND GRAVITATION

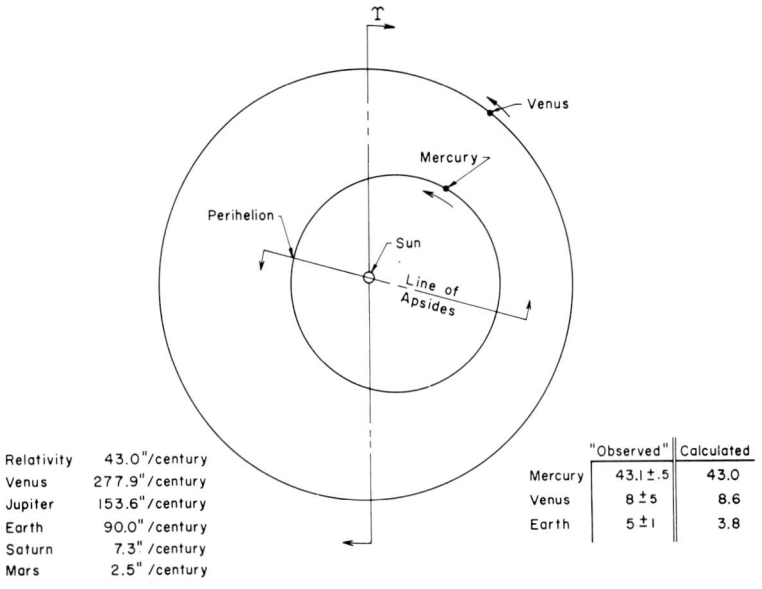

FIG. 14. The perihelion rotation of the planet Mercury. The various contributions are tabulated at the left.

amounting to 43" arc/century,[6] raised a great search during the nineteenth century for possible causes (see figure 14). It had been known from the time of Newton that the planets should move in approximately elliptical orbit about the sun and that, as a result of gravitational interactions with other planets, the major axis of the ellipse should slowly rotate in the same direction as the planet itself. The point of closest approach to the sun is called the perihelion and the rotation of the major axis is usually termed the motion of the perihelion.

In the case of the planet Mercury the gravitational pull of the planet Venus causes the orbit of Mercury to rotate some 278" arc/century. In like fashion the gravitational interaction of the earth on Mercury induces a rotation of

---

[6] G. M. Clemence, *Astr. Papers Am. Ephem.* **11** (1943) : p. 1; *Rev. Mod. Phys.* **19** (1947): p. 361; *Proc. Amer. Phil. Soc.* **93** (1949) : p. 532; R. L. Duncombe, *Astr. Papers Ephem.* **16** (1958) : p. 1; P. A. Wayman, *Quart. Jour. Roy. Astr. Soc.* **7** (1966) : p. 138.

Mercury's perihelion by about 90″ arc/century. The net effect of all the gravitational interactions of the planets on the planet Mercury is to induce a rotation of Mercury's perihelion by about 530″ arc/century, over ten times as much as the interesting relativistic effect. While the relativistic effect is small compared with the effect of these planetary perturbations, the perturbations can be calculated with such accuracy and the observations themselves are so accurate that the excess motion of 43″ arc/century is accurately known, probably with an accuracy of 1″ arc.

It was natural enough during the nineteenth century to attempt to find other bodies, sources of gravitational field, which might be the source of the excess motion. One explanation given for the excess motion evoked Vulcan, a hypothetical planet closer to the sun than Mercury.[7] This planet is still undiscovered, and seeems unlikely ever to be found. In like manner, an attempt made to find significant sources of explanation in the gravitational fields of dust, gas, and meteorites was without success.

Only one of the explanations offered during the nineteenth century seems reasonable today. A distorted sun is capable of inducing part of the perihelion rotation. If the sun were slightly oblate, the resulting distortion of the gravitational field about the sun would induce a motion of the perihelion of the planet. While this possibility had been raised in the nineteenth century, we now know that it is not possible to account for all of the excess motion (43″ arc/century) in this way. A distorted sun would have other effects on the orbit of the planet Mercury, primarily the generation of a slow decrease in its inclination. This effect would be sufficiently large (2.3″ arc/century) to preclude an explanation for the whole of the excess 43″ arc/century as an effect of a distorted sun.

Under General Relativity the whole of the observed excess motion is accounted for as a relativistic effect, but

---

[7] J. Chazy, *La Théorie de la relativité et la mecanique céleste* (Paris, Gauthier-Villars, 1928).

under the scalar-tensor theory the expected relativistic rotation would be [4]: $[1 - (\frac{4}{3})s] \times$ (Einstein's value) or $0.91 \times$ (Einstein's value) $= 39''$ arc/century for $\omega = 5$. Prior to the observation of the solar oblateness, only crude methods were available to estimate ω and this constant had been judged to lie in the range $4 - 8$.[8]

If the scalar-tensor theory were to be correct, part of the excess motion, perhaps $2'' - 4''$ arc/century should be caused by a non-relativistic effect, an ordinary non-relativistic perturbation of the orbit. As will be seen, a distorted sun seems to be the source of such a perturbation, accounting for $4''$ arc/century. This effect on the perihelion should be accompanied by a decrease in the inclination of Mercury's orbit, at a rate of[9] $0.21''$ arc/century. This value is to be compared with the observed anomalous decrease of [10] $0.12 \pm 0.16''$ arc/century.

Tonight our concern is the sun. Is it distorted? If so, by how much? If the sun has a distorted interior with the mass distributed in a non-spherical fashion, what is the source of this distortion? Considering the sun's origin, can we give a reasonable explanation for the existence of a distortion?

It is known from the requirements of static balance that the sun, a great ball of hot gas, should be accurately spherical if all its internal stresses were due to the pressure of the gas. Magnetic and velocity fields provide physically reasonable stresses capable of distorting the sun. While the effect of very strong, deeply buried magnetic fields cannot be ruled out, the field strengths required are very large and one might think that such strong fields would have found their ways to the solar surface. Velocity fields in the form of a rapid rotation of the deep interior of the sun seems a more likely source of distorting stress.[9]

---

[8] R. H. Dicke in *Stellar Evolution*, eds. B. G. Marsden and A. G. W. Cameron (N. Y., Plenum Press, 1966).
[9] R. H. Dicke, *Nature* **202** (1964): p. 432.

A rapid rotation would flatten the mass distribution, which is concentrated almost wholly in the inner 70 per cent by radius. But why should the surface rotate slowly (once every 27 days) if the interior rotates rapidly (once every 2 days)? Under certain conditions, the surface can be slowly rotating while the deep interior rotates rapidly, the result of a magnetic brake acting on the surface. The existence of this magnetic brake, exerted on the sun by the solar wind, is now widely recognized although at the time I wrote my first note on the solar oblateness,[9] the situation was unclear (see figure 15). P. J. E. Peebles and I[9] derived the theory of the solar wind torque, but we lacked good data. The numerical estimate was made before the magnetic field strength in the solar wind had been measured. This field strength was crudely estimated from magnetic field measurements at the sun's surface.[9] Interestingly enough this crude estimate was subsequently found to be fairly good and modern measurements have given a torque differing little

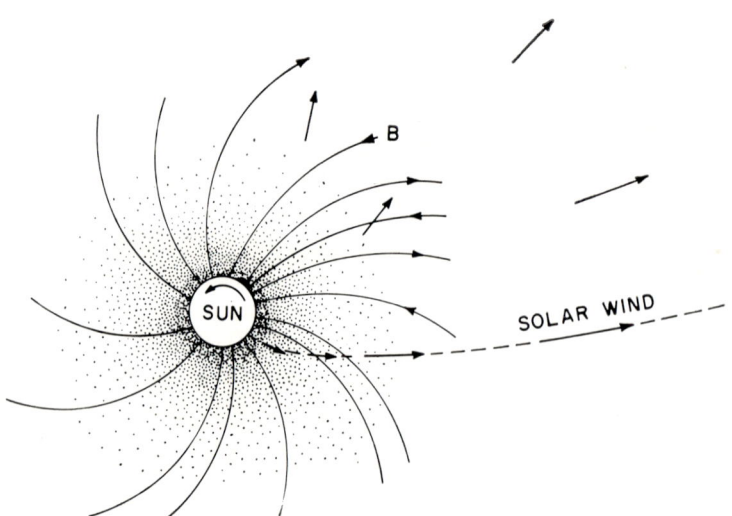

FIG. 15. The solar wind torque. The torque on the sun's surface is caused by the twisted magnetic field lines.

from our original value. Similar theories of the solar wind torque were later independently developed by others.[10]

A possible history for the sun would show the sun and other rapidly rotating stars being formed through the condensation of a very large turbulent gas cloud. The angular momentum of the turbulent gas would be incorporated in the stars formed out of the gas, in part in the rotation of the stars.

Rapid rotation is observed in the stars more massive than 1.6 solar masses. These stars are somewhat bluer than the sun and have surface temperatures in excess of 6,500° K. Rotation is not normally observed in stars redder than this, although rotation is observed in very young red stars.

A very interesting explanation exists for the difference in rotation of the blue and red stars, a distinction brought out by R. Kraft.[11] According to the theory of stellar atmospheres, stars with surface temperatures below 6,500° K. have thick convective layers at their surfaces. This convective zone is turbulent because of heat being carried convectively to the surface of the star, and this continual boiling motion of the surface radiates sonic waves into the upper atmosphere of the star, heating the atmosphere. The continuous heating of the upper atmosphere evaporates it and drives a stellar wind. The picture evoked is one similar to the sun. A red star is believed to be acted on by a stellar wind which tends to slow down the rotation of its surface through a torque which results from a twisting of the magnetic field trapped in the wind. A blue star is believed not to possess such a convective layer, hence no stellar wind or surface torque.[11]

By making observations on clusters of stars, R. Kraft has shown that red stars found in young clusters do show rotation and that the younger the cluster the more rapid is

---

[10] E. J. Weber and L. Davis, Jr., *Astrophys. Jour.* 148 (1967): p. 217; J. L. Modisette, *Jour. Geophys. Res.* 72 (1967): p. 1521; A. Alfonso-Faus, *Jour. Geophys. Res.* 72 (1967): p. 5576.

[11] R. Kraft, private communication (1964).

the rotation.[12] It is possible to account nicely for the slowing down of the rotation of these stars if one assumes a torque acting on the surface of the star comparable with our known solar wind torque and if it is only the outer layers of the star that are slowed down, not the whole star. Alternatively, if the torque were fifty times as great, the whole star could be slowed in its rotation.

Under certain conditions the interior of the sun should be capable of spinning faster than the surface, freely and without turbulence, the angular momentum leaking slowly to the surface through the effect of laminar viscosity. The sun is so enormous that this mechanism for losing angular momentum is relatively ineffective and even a weak torque applied to the stellar surface is capable of keeping the surface slowly rotating.

The picture that we would paint is the following: the sun when first formed was a rapidly rotating ball of gas driving the solar wind which twisted the magnetic field in the solar wind and exerted a torque on the surface of the sun to slow the rotation of the outer 30 per cent (see figure 15). The interior of the sun is pictured as continuing to rotate rapidly, with perhaps a 2-day period. It is estimated that the outside of the sun was slowed by the solar wind torque to near its present rotation period, 27 days, in a few hundred million years.

A test for a flattened gravitational potential distribution is easily obtained if the surface of the sun is sufficiently free of disturbing forces. It is easily proved that the surface of the sun accurately mirrors the shape of the total potential distribution, gravitational plus centrifugal, in a coordinate system rotating with the average angular velocity of the solar surface, providing that velocity and magnetic fields at the surface of the sun are sufficiently small,[9] as viewed in the rotating coordinate system. We are faced with two questions at this point: First we can ask whether the surface

---
[12] R. Kraft, private communication (1968).

of the sun is slightly oblate; second we can ask whether an oblate sun, assuming that oblateness is observed, does in fact properly mirror a flattened gravitational potential, the effect of a distortion of the deep interior of the sun. In my opinion the answer to both of those questions is yes.

Consider the first problem of the observation of the solar oblateness. The expected distortion is extremely small, only 5×10^{-5} for $\omega = 5$, i.e. $(r_{eq} - r_p)/ = 5 \times 10^{-5}$. This small oblateness represents a difference of only 35 kilometers between the radius of the sun at the equator and that at the pole. Stated in another way, as seen from the earth these radial distances differ by only 0.05" arc. Unfortunately, it is necessary to observe the sun during the day when the air is highly turbulent. This turbulence causes a blurring of a telescope image amounting to some 3" arc, far larger than the looked-for 0.05" arc. It is not even possible to use the standard solar astronomer's trick of rising early to get to the telescope before the sun is so high in the sky as to make the air warm and highly turbulent. We found it necessary to make our observations within a few hours of noon in order for the sun to be sufficiently high in the sky for the atmospheric refractive distortion of the sun's image to be manageable.

My colleagues H. Hill and H. Mark Goldenberg and I designed and built a specialized telescope to observe the solar oblateness. The instrument employed a technique seldom used by astronomers but capable of eliminating most of the difficulties due to poor daytime seeing. With the help of a number of students, R. Stokes, P. Henry, K. Davis, E. McDonald, and D. Hawley, Goldenberg and I carried out a series of measurements with this instrument during the summers of 1966-1968.[13]

The principle of this instrument is shown in figure 16. An image of the sun is projected on an opaque occulting disc permitting the light from a narrow annulus at the

---

[13] R. H. Dicke and H. Mark Goldenberg, *Phys. Rev. Letters* **18** (1967): p. 313.

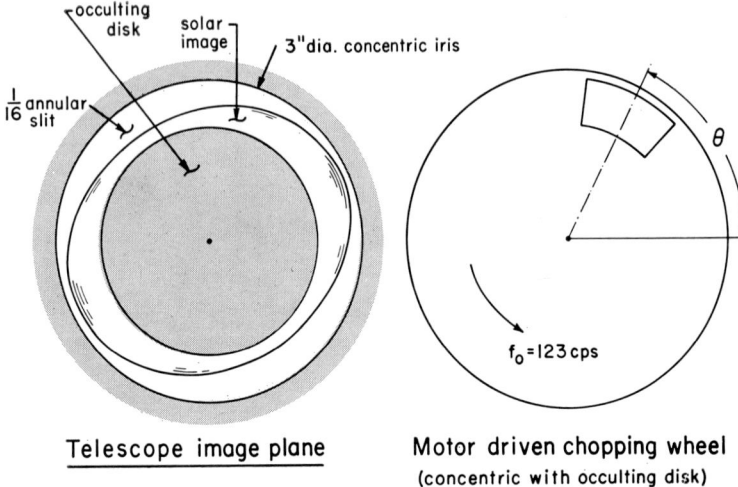

Fig. 16. Image of sun, oblateness exaggerated, projected upon the occulting disk. Simplified view of scanning wheel removed from its normal position behind the occulting disk. Only one scanning aperture is shown in the wheel whereas two diametrically opposed apertures of slightly different sizes were used.

outside of the sun's disc to pass the occulting disc. This light falls on a rapidly rotating wheel carrying two unequal apertures. These oppositely situated apertures rapidly scan the edge of the sun determining whether the light flux is uniformly distributed about the occulting disc. The situation with a greatly exaggerated solar oblateness is shown in the left side of this figure. In figure 17 we see curves representing the variation in light flux as a function of time for the two cases, of an oblate sun, and a displaced sun.

In figure 18 is shown the elements of the optical system.[13] Light from the sun is reflected from a very flat quartz primary mirror which is so mounted and motor-driven as to follow the sun and send the sun's light to a secondary mirror. Light from the secondary mirror passes down a vertical axis through an objective and other lenses of a telescope to form an image of the sun on the occulting disc shown just above the scanning motor. The occulting disc is pierced by a small hole in its center permitting a sample

SOLAR OBLATENESS AND GRAVITATION

of the light from the center of the solar disc to pass to a photocell which serves to monitor the haze condition in the sky. The remainder of the light passing the edge of the occulting disc and passing through the scanning wheel is brought to focus on a second photocell. It is this second photocell which provides the information concerning the centering of the solar image and the oblateness of the solar image.

Information regarding the centering of the image is fed back electronically to determine the position of the primary mirror. This feed-back mechanism is capable of keeping the sun's image very accurately centered on the occulting disc. The oblateness signal is processed electronically, averaged, and recorded for analysis. (See figure 19.)

The vertically mounted telescope permits the elimination of gravitationally induced errors in the telescope. The telescope, from the objective down to the photocell at the bottom, is so mounted as to be rotated about a vertical axis.

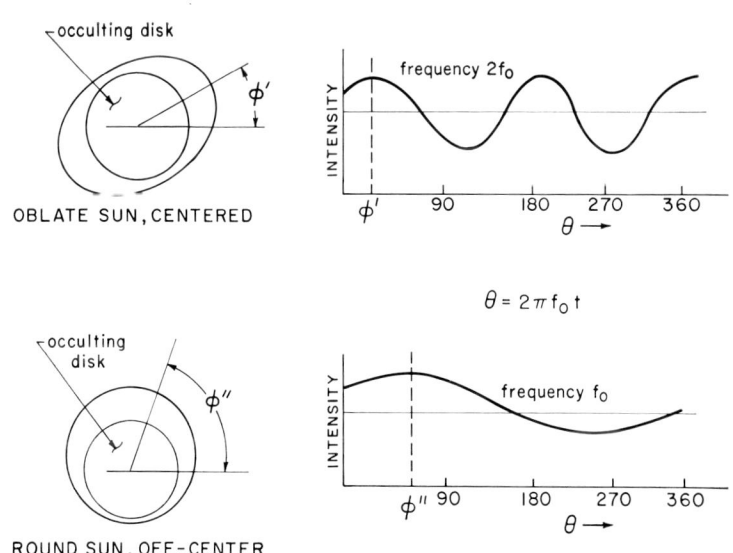

FIG. 17. Graphs showing the angle (and time) dependence of the photoelectric current derived respectively from an oblate solar image, and from a displaced solar image.

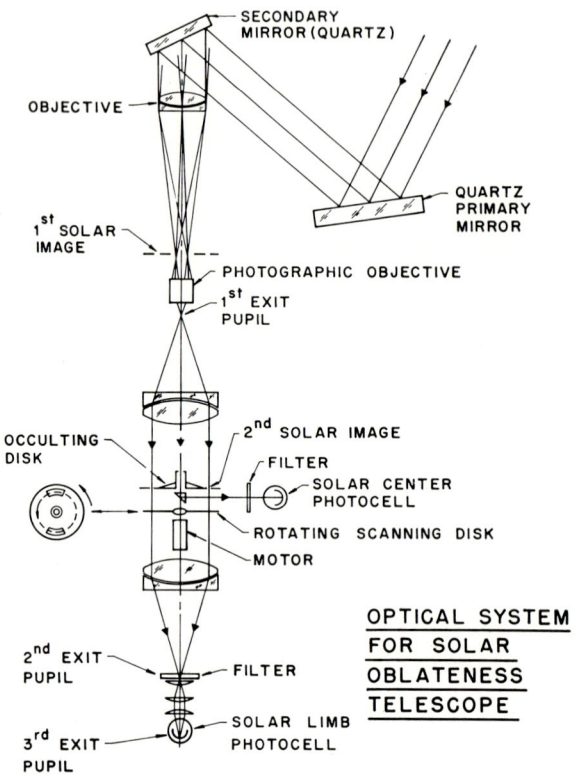

Fig. 18. The optical system of the instrument used to measure the solar oblateness. The vertically mounted telescope, from the "objective" down to "solar-limb photocell," is mounted as a single unit and can be rotated about the vertical optical axis.

Typically, the oblateness is averaged electronically for a minute and the telescope is then rotated through 90° and the observation repeated for a second minute. The average of these two measures is free from telescope errors. In like manner, errors introduced by astigmatism to the two mirrors are eliminated by rotating these mirrors 90° about axes perpendicular to their surfaces after each day of observation the change of one mirror occurring at noon, the other at day's end. Equal numbers of observations were taken with the two mirrors in all four possible configurations.

In so far as we could see, all significant instrumental errors were removed by these processes except for one error involving the primary quartz mirror. An off-axis astigmatic error is associated with a slight spherical shape of the quartz mirror. Owing to the proximity of the secondary mirror to the objective, this effect is small for this mirror. This small effect averages to zero through the rotation of the stellar image relative to the secondary mirror. The

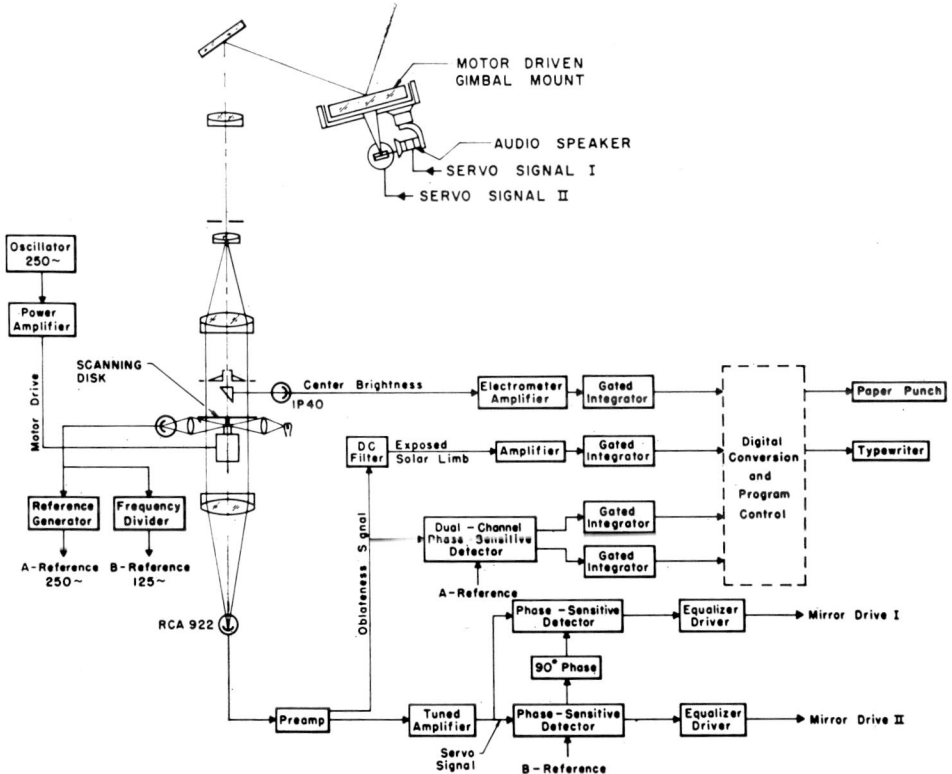

FIG. 19. Block diagram of the instrument, showing the main electronic components in relation to the optical system exhibited in figure 18. Note that the output lines from the "equalizer drivers" at the bottom of the figure are to be connected to the "servo signal" inputs at the primary mirror. These control voltages actuate audio speaker-coil assemblies to bend flexible linkages supporting the mirror-mount. Crude pointing of this mirror is derived from a synchronous motor-drive, but the fine pointing is derived from the servo-controlled bending of the flexible hinge system.

effect is larger for the primary mirror and the image is fixed in this mirror. The off-axis astigmatic effect could not be eliminated, but it could be bypassed.

This distortion is confined to the north-south line in the sun's image. The oblateness can be expressed as a combination of a shortening along this line and a shortening of the image along the northeast-southwest line. The component of the oblateness along this latter line, called the diagonal component of the oblateness, is completely free of the off-axis error.[13]

Figure 20 shows the mirror mount and mirror for the large primary quartz mirror and also the mirror mount for the small secondary mirror. The opening is into the roof of the very small building built for the purpose of housing this instrument. Figure 21 shows the interior of the building with three relay-racks full of electronic equipment and

FIG. 20. The upper end of the fixed telescope housing, showing the separately mounted primary mirror. At the top of the fixed telescope support housing can be seen the back of the secondary mirror-mount.

SOLAR OBLATENESS AND GRAVITATION

Fig. 21. The interior of the compact "instrument shack" in which were housed the telescope and the electronic equipment. Note the thin rods projecting from the upper part of the telescope tube. These serve as handles to rotate the tube. The "black stripe" on the tube shows the location of an access door into the tube. This door permits the placement of various occulting and calibration disks in the tube.

the telescope itself at the far end of the building. I would guess that the ratio of telescope aperture to pounds of electronic equipment is far smaller with this instrument than any other modern telescope. The telescope itself had an aperture of only 2 inches.

By integrating the total light flux passing the occulting disc and scanning wheel, the main difficulties associated with poor daytime seeing were eliminated. We would have been unable to do nearly as well attempting to measure directly the position of the ill-defined limb of the sun. A second factor essential for the success of the observational program is the fact that the axis of the sun, as it is seen in the sky during the summer months, changes rather rapidly (figure 22). This effect has nothing to do with the motion of the sun's rotational axis, but rather it is the effect of our moving around the sun and observing the rotation axis from different directions. On July 7 the sun's rotational axis is in the north-south direction and it is at that time that the orientation is changing most rapidly.

Averaging the times of the observations, the distortion of the sun should be along its rotation axis. Consequently, we have a known and expected time dependence for the orientation of the solar distortion. Without this expected variation with time it would be very difficult to support the interpretation of the observed small signal as the effect of a solar oblateness rather than some instrumental atmospheric effect. But on the contrary, the variation of the signal through the summer indicated that the distortion was

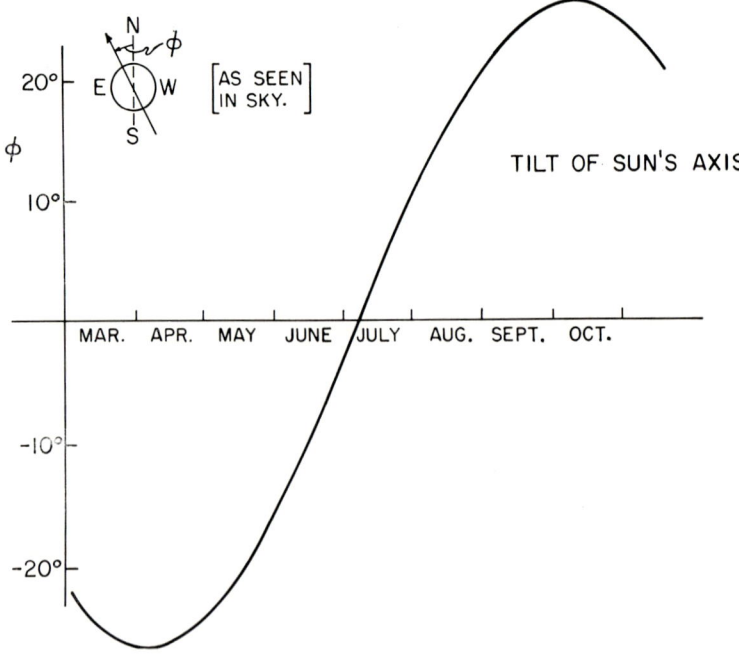

FIG. 22. The tilt of the sun's axis of rotation.

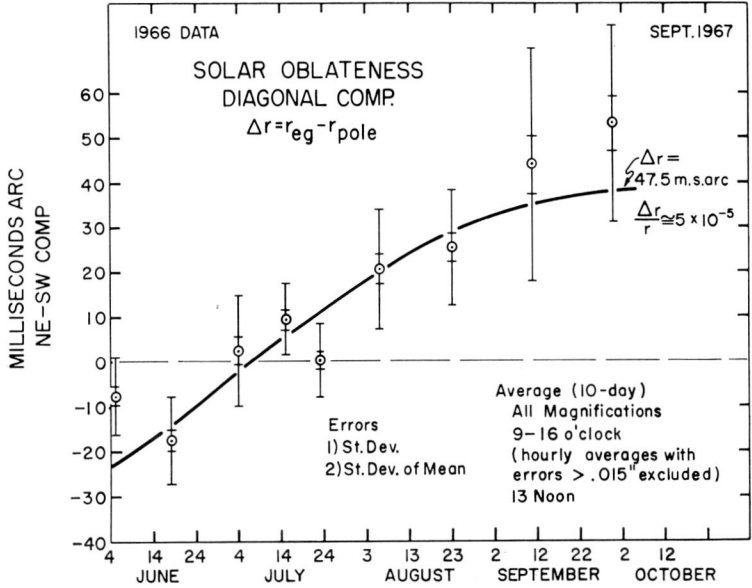

FIG. 23. The "diagonal component" of the sun's oblateness, the data of 1966. The curve is computed from figure 22, assuming that the sun has an oblateness $\Delta r/r = 5 \times 10^{-5}$.

along these same lines as the sun's rotation axis within 1 or 2 degrees. The distortion axis appeared to follow the rotational axis in the sky while the rotation axis precessed through an angle of 40°. I believe that a distortion of the sun was observed, not some peculiar instrumental or atmospheric effect.

Figure 22 shows the orientation of the sun's rotational axis in the sky through the summer months.

Figure 23 shows a curve derived from figure 22 showing the expected variation of the diagonal component of the oblateness with time for an oblateness of 5×10^{-5} and the measurements made during this period of time. The measurements in question constitute the averages of data taken during 1966 over approximately 10-day intervals, between the hours of 9 and 16 o'clock, with noon at 1300.

These observations were made with three different amounts of the sun protruding beyond the occulting disk, the amounts being 19.2″, 12.9″, and 6.3″ arc. This was done to permit the separation of two different effects, the contribution from a brightness variation about the limb of the sun and the solar oblateness itself. Two different limb exposures would have been enough but three were used to provide an internal check on the method. It is clearly important to be sure that the signal being observed is due to solar oblateness and not to a variation with latitude in the brightness of the solar limb, but the brightness distribution is also needed for another purpose to which I shall return.

Figure 23 shows data points representing averages over all three magnifications but figure 24 shows the results broken down into separate data points for each of the three

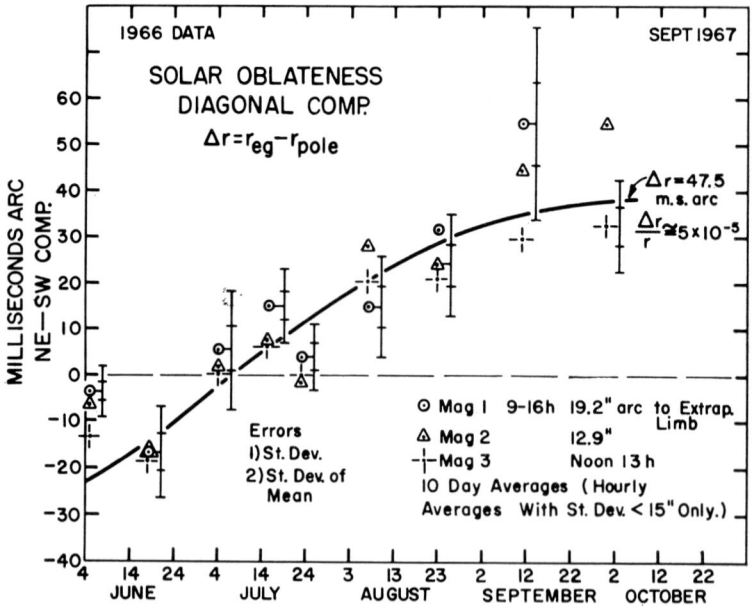

FIG. 24. The same data displayed in figure 23 but broken down into averages over the three different magnifications of the solar image. This provides measurements obtained with three different amounts of the solar disk projected beyond the occulting disk. Observations with all three degrees of exposed solar limb (19.2″, 12.9″, and 6.3″ arc) permit the separation of the signal into two parts, one due to the oblateness and one due to a hypothetical brightness difference between the equator and pole.

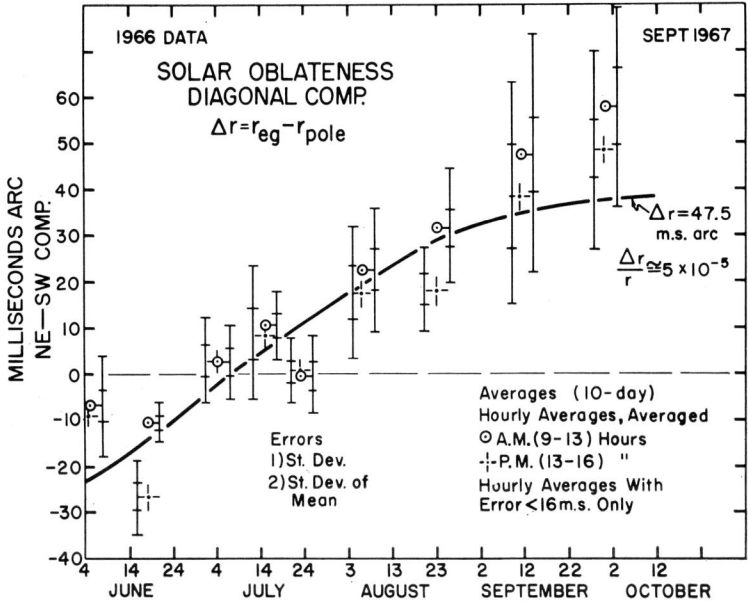

FIG. 25. The 1966 data broken into morning and afternoon groups as a test for a time-dependent systematic error.

magnifications. Figure 25 shows a similar breakdown with the data divided into morning and afternoon hours of observation. Figure 25 provides a test for a possible effect associated with the position of the sun with respect to Princeton.

Figures 23-25 represent the data taken in 1966 analyzed in several different ways.[13] The data taken in the following year are consistent with the above 1966 data. The net conclusion obtained from these measurements is that the sun has an oblateness of 5×10^{-5}, that is, the equatorial radius of the sun exceeds the polar radius by 50 parts in a million. It is also concluded from these measurements, and the ones made in 1968, that the sun is remarkably uniformly bright in the mean. This is not to ignore the tiny little blotches, sun spots and faculae, but that averaged over the season there is no reliable determination of a net brightness difference between the pole and the equator. This latter observation is of considerable importance when it comes to the interpretation of the observations.

As remarked above, the shape of the sun is determined by the shape of a surface of gravitational potential when there are negligible distortions induced by surface fields.[9] On the other hand, one knows that magnetic and velocity fields do exist in the surface layers. It is necessary to inquire about their significance. It is possible to show that a distortion of the surface layers of the sun induced by surface magnetic or velocity fields is also normally accompanied by a distortion in the temperature distribution over the surface. Only for a very select class of surface fields is it possible to generate a distortion of the surface without at the same time generating an impossibly large temperature distribution over the surface.[14] In this way we are limited to a restricted range of surface fields if the condition that the sun be uniformly bright is to be satisfied. This restricted class of surface fields has been looked at with considerable care with the conclusion that the fields observed at the solar surface are incapable of generating the oblateness and that the solar oblateness is associated with a distortion of the deep interior of the sun. The observations and their interpretation are delicate, and it is not possible to be absolutely certain about this conclusion, but I have considerable confidence in it. I would find it quite difficult to believe that the observations have been incorrectly interpreted, but it is a separate matter when one considers the connection between the gravitational structure and the oblateness of the sun. We have been unable to find reasonable surface-field distributions which would produce this distortion, and it is my belief that the observed oblateness does represent a true gravitational oblateness of the sun. However, it would be a mistake to be dogmatic about this; the question is a very difficult one.

It has been argued by Howard, Moore, and Spiegel[15] that

---

[14] R. H. Dicke, to be published.
[15] L. N. Howard, D. W. Moore, and E. A. Spiegel, *Nature* **214** (1967) : p. 1297.

the sun cannot possess a rapidly spinning core, that angular would be quickly transferred to the outside by "spin-down," the convection of angular momentum by Ekman currents responsible for the rapid slowing of rotation in a cup of tea. I have argued that Ekman currents are forced only when pressure and density are functionally connected, not true of the deep solar interior.[16] Thermally driven circulation currents will generally occur, but these currents are usually very slow and their magnitudes depend upon the details of the distribution of angular velocity in the solar interior.

Goldreich and Schubert[17] have argued that a rapidly rotating core would loose angular momentum by an analogue of the "salt finger" instability familiar to oceanographers. Their instability is associated with the detachment from the rapidly rotating of thin toroidal rings of fluid which carry angular momentum upward. Their analysis is carefully carried out but the set of conditions needed in the solar interior may not be justified. The assumed absence of meridional motion, a deeply buried magnetic field, or a compositional gradient from the zone of differential rotation is questionable.

If it be assumed that the sun is distorted in a way indicated by the solar oblateness several conclusions can be drawn. The first is that roughly 4" arc/century of the excess motion of the perihelion are due to the oblateness of the sun. This leaves roughly 39" arc/century to be explained as a relativistic effect. This is in good agreement with expectations under the scalar-tensor theory but is in disagreement with the expectations of ordinarily unmodified general relativity. It should be remarked that there is some uncertainty regarding the orbit of Mercury. These observations are difficult and it is conceivable that the errors in these observations are greater than astronomers

---

[16] R. H. Dicke, *Astrophys. Jour.* **149** (1967): p. L121.

[17] P. Goldreich and G. Schubert, *Astrophys. Jour.* **150** (1967): p. 571; **154** (1968): p. 1005.

presently believe. In due course the planetary radar observations will provide an independent measure of the perihelion rotation of Mercury.

A second conclusion drawn from the oblateness observations, assuming that the rapidly rotating core explanation is correct, concerns the sun as a rotating star. R. Kraft has shown for blue stars that the spin momentum varies with stellar mass as the 1.57 power of the mass.[11] Assuming that the sun has a core rotating with a 2-day period, the sun's spin angular momentum would be consistent with Kraft's relation. A reasonable generalization would be to assume that Kraft's relation holds for all normal main sequence stars, but that the angular momentum is deeply buried in red stars. Thus a significant difference between red and blue stars is to be found in the stellar wind that slows the surface rotation of a red star.

What lies in the future? First of all, in future years the planetary radar observations should be capable of giving us better independent measures of the rotation of the perihelion of the planet Mercury. These observations should also be capable of giving us perihelion rotation measures of both the earth and Mars. By combining data from several planets it is possible to eliminate the effect of an oblate sun and hence to check for the relativistic effect directly. It is also possible that the other orbital effects associated with an oblate sun, perhaps for an artificial planetoid or an asteroid tagged with an electronic system, will eventually provide new independent determinations of the gravitational distortion of the sun. Finally it must be remarked that, if our observations are interpreted to favor the scalar-tensor theory over general relativity, this interpretation implies a gravitational deflection for light only 0.93 of Einstein's value. The improved observations possible with planetary radar, and the other ways of improving the determination of the deflection of light, may provide results which will tend either to confirm or to discredit this interpretation of our observations of the oblateness of the sun.

III. The Cosmic Fireball and Gravitation

Introduction

In the previous two lectures it had been emphasized that gravitation is extremely weak and that large bodies are required for meaningful experimentation. The sun is the source of the gravitational field for the Eötvös experiment discussed in the first lecture and also for the positive tests of General Relativity discussed in the second lecture. Tonight we shall be concerned with the largest dynamical system possible, the whole Universe. Here distance and time spans are so great that the dynamical behavior of the system is dominated by relativistic requirements rather than relativistic effects, being minor perturbation on a system which is almost completely described by Newtonian mechanics.

It is amusing that the whole Universe can be considered an apparatus, but experimentation with such an apparatus has limitations. The observer has no control over the experiment, which cannot be repeated. Furthermore, it is not possible to modify the apparatus.

It is not only the study of gravitation that requires apparatus on the astronomical scale. Nuclear reactions are occurring in the interior of the sun under conditions which cannot be duplicated in the laboratory, and in the interior of white dwarf stars and neutron stars material is compressed to densities far greater than can be obtained in the laboratory. Magnetohydrodynamic effects involving magnetic fields and very low density ionized gas are occurring in the Universe on distance and time scales far greater than anything we can duplicate in the laboratory. Astronomical systems are essential to the investigation of such physical phenomena. But, it is in connection with gravitation, the weakest of all known interactions, that these astrophysical approaches to experimental science are essential.

The Universe is far too complicated a structure to be studied deductively, starting from initial conditions and solving the equations of motion. It is much more productive to examine with care the observable properties of the Universe and then attempt to draw conclusions concerning its nature and past history.

What do we see when we look out into the Universe? Except for the planets, moon, and other tiny fragments of the solar system, the first significant object encountered is our own sun. The sun, a typical star, contains in addition to hydrogen, roughly 25 per cent helium and 2 per cent other elements in its outer layers. It generates its radiation by converting hydrogen into helium at very high temperatures deep in its interior. The fact that it contains elements other than hydrogen and helium is very significant, for there are reasons to believe that these heavier elements were produced earlier in the interiors of other stars.

Leaving the sun behind and journeying much farther into space, over 100,000 times as far as the distance to the sun, we begin to reach other sunlike objects, the stars. Stars sometimes occur in clusters, gravitationally bound systems, such as the well-known Pleiades, and the faintest, hence most distant, stars tend to be concentrated in a great band across the sky, the Milky Way. This latter structure suggests a most striking thing about the distribution of stars, that they tend to occur in a flattened, disk-shaped structure. It is this disk seen edge-on that represents the Milky Way. A careful study of this disk structure, particularly with the aid of the radio-telescope, has shown that it is in the form of a flattened spiral and that it has a form similar to the long-known spiral nebulae which occur in large numbers in the sky. This disk is a gravitationally bound collection of stars, the Galaxy. It is known that the Galaxy is approximately 25 kiloparsecs in diameter,

7.5×10^{22} cms or 5 billion times the distance to the sun. It contains roughly 10 billion stars and the thickness of the disk near the sun is roughly 2 per cent of the diameter. The whole disk is slowly spinning about its axis, the outer parts more slowly, and the sun takes 250 million years to revolve about the Galaxy.

One aspect of the structure of the Galaxy not obvious to the eye is the existence of two different populations of stars, a first population, or Population I, characterized by the above-mentioned flattened-disc-type distribution, and a spherical second population, Population II, having a distribution about the center of the Galaxy. Population II stars do not move in the circular orbits characterizing the Population I stars, but in highly eccentric orbits. The Population II stars are known to be extremely old and to be relatively free of the heavy element found in the sun. The differences in composition and motion are of significance to a fundamental question involving gravitation, a question to which we shall return.

The ages of stars are inferred from studies of clusters of stars. It is believed on both observational and theoretical grounds that a gravitationally bound cluster of stars consists of stars essentially the same age condensed from a gravitationally collapsing cloud of gas. The stars have various masses and the more massive are the more luminous, the luminosity increasing as the cube of the mass for stars more massive than the sun. Thus the lifetime of such a star, before exhausting hydrogen at its center varies roughly inversely as the mass squared. Thus the maximum luminosity of the stars still burning hydrogen at the center provides a measure of the age of the cluster. The difficulty in assessing the maximum luminosity is ameliorated by making use of a relation between luminosity and color. (The most massive and luminous stars are the bluest.) It is only necessary to observe the colors of the stars along the "main sequence," stars burning hydrogen at their centers, to obtain an estimate of the age.[2]

The central facts which emerge from a study of stellar ages is that Population II stars are probably all of about the same age, whereas the Population I stars are of all ages younger than the Population II. The absolute ages are uncertain, particularly those of Population II stars, because of doubt concerning composition. In principle, the theory of stellar evolution provides an age for a cluster of stars, but only after the composition is known. Depending upon assumptions, the age of Population II stars obtained in this way ranges from 9×10^9 years to 20×10^9 years.[3] The younger age corresponding to the most helium in the stellar mixture. The age of the Galaxy can also be inferred from the radioactive decay of the uranium produced in the young Galaxy.[1] This age, derived from the most accurate of the ways of dating the Universe, is 7×10^9 years.[4] We shall return to a third way of inferring an age.

The great spherical distribution of Population II stars is called the halo. It contains approximately 100 globular clusters. These clusters, spherical in shape and similar in size and luminosity, contain approximately 10^5 stars (see fig. 26). The compositions and ages seem to be the same as the individual, or field, Population II stars. P. J. E. Peebles and I believe that these clusters represent the fossil remains of the first objects condensed out of the expanding Universe. But the expansion of the Universe is another question to which I must now turn.

When we look beyond our own Galaxy, we encounter many other very large collections of stars, in many cases disk-shaped and containing spiral arms similar to those in our Galaxy.

---

[1] E. M. Burbidge, G. R. Burbidge, W. A. Fowler, and F. Hoyle, *Rev. Mod. Phys.* **29** (1957): p. 547.

[2] M. Schwarzschild, *Structure and Evolution of the Stars* (Princeton, N. J., Princeton University Press, 1958).

[3] I. Iben and J. Faulkner, *Astrophys. Jour.* **153** (1968): p. 101.

[4] R. H. Dicke, *Astrophys. Jour.* **155** (1968): p. 123.

Fig. 26. Globular cluster. Some of us believe that these compact clusters of roughly 10^5 stars are fossil remains of the first gas clouds condensed from the cooling cosmic fireball.

Before the advent of photographic techniques in astronomy it had not been generally appreciated that the faint wisps of light with a vaguely spiral form were vast collections of stars. Even now, only the brightest stars in the nearest of the spiral nebulae are individually detected.

We now recognize that our Galaxy is only one of perhaps a billion similar gravitationally bound systems of stars in the Universe. Not all of the stellar collections are spiral galaxies. In some cases the galaxies have a smooth, elliptical shape and sometimes consist of a bar trailing spiral arms. There are also highly irregular galaxies such as the famous Magellanic clouds seen in the southern hemisphere.

These types of galaxies are all found at various distances, enormous numbers at great distance. For each type of galaxy there is not the uniformity in size and luminosity characterizing the globular clusters. The masses and luminosities of galaxies range over four orders of magnitudes.

When the spectral lines of distant galaxies are observed, it is found that they are shifted toward the red, a Doppler effect representing a motion away from us. This so-called galactic red shift is found to be proportional to the distance of the galaxy. The connection between red shift and distance has been investigated with great care by A. Sandage who has found the most useful tool for this investigation to be the cluster of galaxies. Clusters of galaxies occur in large numbers in the sky. These clusters appear to be gravitationally bound systems of galaxies with internal motions of about 10^3 km/sec. Sandage has found that the brightest galaxy in a cluster tends to have a rather standard brightness. The distance to the galaxy is inferred from the apparent luminosity of the brightest member of the cluster. Thus a relation between distance and red shift is obtained.[5]

The observations show that the Universe is an expanding structure, with an expansion rate proportional to the distance from us. This relation between distance and velocity is highly significant. In 1929 Hubble discovered that the Universe is exploding, with the distance of travel proportional to the outward velocity.[6] If it be assumed that the outward velocity of a galaxy does not change with time, it is a simple matter to compute the time at which this Galaxy was very close to us. This time, the same for all galaxies, is presently believed to be 10-17 billion years.[5] On the other hand, the gravitational pull of other galaxies tends to slow down the expansion rate and, as will be discussed, there are some reasons to believe that the true age of the Universe, the time back to the explosion, is approximately two-thirds of the above value or 6.5-11 billion

---

[5] A. Sandage, *Astrophys. Jour.* **152** (1968): p. L149.
[6] E. P. Hubble, *Proc. Nat. Acad. Sci.* **15** (1929): p. 168.

years. It should be noted that the uranium-dated age of the galaxy, 7 billion years, is consistent with the above poorly determined age.

Particularly significant in the distribution of galaxies about us is uniformity and isotropy. The galaxies appear to be uniformly distributed about us. Not only is the distribution uniform but the above described motions with respect to us represent a uniform dilation. How is this to be interpreted? We might be tempted to conclude that man occupies some special central point in the Universe, that galaxies move away from *us*. An alternative interpretation is that the Universe is uniform in structure and that all points are similar. Thus the Universe might appear isotropic from any particular galaxy in which man happened to be living. This latter interpretation seems to be consistent with the observations. It is a simple matter conceptionally to transport ourselves to another distant galaxy, if we take into consideration the distributions of galaxies, in so far as they are known, and their motions (assumed to be purely radial) relative to us. The mathematical transformation is easily carried out and leads to the conclusion that in the average the Universe would appear the same when seen from other galaxies. This is consistent with the assumption that the Universe is uniform and that man does not occupy a preferred central galaxy.

The overall picture which emerges is one of a Universe born in a catastrophe and perhaps heading toward a catastrophe. If the Universe is slowing sufficiently in its expansion rate, it will some day stop expanding and start contracting. This could lead to a catastrophe with galaxies colliding with each other. The end of the Universe would then be similar to its beginning, our main concern tonight.

Gravitational Theory and the Expanding Universe

As was discussed in the two previous lectures, we are intercomparing two relativistic theories of gravitation, Einstein's General Relativity and a closely related general relativistic theory, the scalar-tensor theory. Both are based on Riemann's view of geometry, a physical space which is not necessarily flat (Euclidian). In Einstein's gravitational theory gravitational effects are associated with space curvature. In the scalar-tensor theory the point of view is similar, but a scalar field also plays a role, a role similar to that of the scalar field of pre-relativity gravitational theory.

For both of these theories, space is reasonably flat over the small four-dimensional space-time volume required for a laboratory experiment. Newton's view of dynamics in an absolute, flat, three-dimensional space is adequate in such a small coordinate patch, if particles move sufficiently slowly relative to the laboratory. Rapid motions require Einstein's special relativity. This is based on a flat four-dimensional space-time.

The whole Universe is so enormous that such a limited view based on the notion that a small element of space-time is reasonably flat would be expected to be adequate for discussing only a fragment of the Universe. But, owing to the uniformity of the Universe, any small space-time volume-element contains, in the mean, a representative sample of the whole. The sample should be small, but not too small. It is essential that it be large enough to contain many galaxies, even clusters of galaxies, to permit the interpretation of the matter as a smooth, uniform, continuous fluid.

Fortunately, the whole Universe is so enormous that even so large a volume as this can be small by comparison with the Universe. Thus, we are fortunate in being able to study the whole Universe by examining a representative sample contained in a sensibly flat volume-element.

A striking feature of Newton's gravitational theory is the vanishing of all gravitational effects everywhere inside a hollow spherical mass shell. Intuitively one might expect a test body close to the shell to be attracted to the nearest

THE COSMIC FIREBALL AND GRAVITATION 57

point of the shell. This does not occur, for little matter is contained in this limited region of proximity whereas the opposite and distant part of the shell contains much more matter. The two oppositely directed attractive forces just balance. If a hole were drilled to the center of a uniform earth, the acceleration of a pellet dropped in the hole would keep decreasing as the pellet passed inside more and more of the spherical layers of the earth.

That gravitational forces vanish in the interior of a spherical cavity in an isotropic mass distribution holds also for General Relativity. In the interior, space-time is flat and there are no gravitational effects.

Such a large spherical cavity does not exist in the Universe, but we can produce it in our imagination—scooping the galaxies out of a large, but not too large, spherical cavity and then replacing them one by one in the resulting flat-spaced cavity. In replacing the galaxies we include their gravitational effects within the spirit of Newtonian gravitation, as gravitational attractions between the various galaxies contained in the spherical hole. This picture, all galaxies lying inside an imaginary spherical volume interacting gravitationally with each other but ignoring the outside matter, permits a simple description of the expansion of the Universe.[7]

Imagine a large spherical volume in space with ourselves at the center. The radius of the sphere must be small compared with the radius of the "visible Universe," but it should be large enough to contain a very large number of galaxies. A galaxy lying inside this volume element but near the spherical surface would be acted on by a gravitational force directed toward the center as though all the matter in the spherical volume were concentrated at the center. The resulting acceleration of this test object, the galaxy, toward the center, or the deceleration of the rate

---

[7] W. H. McCrea and E. A. Milne, *Quart. Jour. Math.* 5 (1934): p. 73; C. Callan, R H. Dicke, and P. J. E. Peebles, *Amer. Jour. Phys.* 33 (1965): p. 105.

of expansion of the Universe, would be

$$a = \frac{GM}{r^2} = \frac{4\pi}{3} G\rho r, \qquad (1)$$

where M is the mass contained in the sphere, r is its radius, and ρ represents the average matter density of the Universe. G is the gravitational constant.

From the uniformity of the expansion of the Universe, the radial velocity of this galaxy away from us is

$$v = T_h^{-1} r, \qquad (2)$$

where T_h, called the Hubble age of the Universe, is named after the discoverer of the expansion. T_h would be the age of the Universe if the expansion occurred without deceleration.

The magnitude of the deceleration is conveniently characterized by the dimensionless deceleration parameter

$$\frac{ar}{v^2} = \frac{4\pi}{3} G\rho T_h^2 . \qquad (3)$$

Remembering that $a = -d^2r/dt^2$ and that ρr^3 is a measure of the matter contained in the sphere and is independent of time, eq. (1) can be integrated to give

$$v^2 = \frac{8\pi}{3} G\rho r^2 + v_o^2, \qquad (4)$$

where v_o^2 is a constant of integration. If the constant is positive the Universe expands without limit, but for a negative value of v_o^2 the Universe expands to a maximum size and then contracts again. For $v_o^2 = 0$, the Universe expands without limit but continuously slows in its rate of expansion.

Equation (4), from a crude non-relativistic point of view, is a statement of energy conservation. After multiplying by ½ ρ and transferring the first term on the right to the left side, the three terms are respectively, per unit volume, the kinetic energy, the potential energy, and the total energy.

While eq. (4) was derived under the assumption of zero pressure, i.e. the pressure associated with the random motion of galaxies is negligible, it holds also when pressure is present. Remembering Einstein's mass-energy equivalence, the mass density ρ must then include a contribution from thermal energy. For the actual Universe, this energy is believed to be in the form of black-body radiation at a temperature of 2.7° K. This contribution to energy density is presently negligibly small but, as will be discussed, it was very important in the past.

When there is non-vanishing radiation pressure in the Universe the total energy contained in the dilating sphere defined above is not constant because the internal pressure is performing work on the surface of the expanding sphere. The rate of change of the total energy contained in the sphere satisfies

$$\frac{d}{dt}(\rho c^2 r^3) = -P \frac{d}{dt}(r^3). \tag{5}$$

Differentiating eq. (4) with respect to time and substituting (5) gives

$$a = \frac{4\pi}{3} G \left(\rho + 3 \frac{P}{c^2}\right) r. \tag{6}$$

The present average density of the Universe is poorly known. If we believe that the matter resides only in seen matter, the galaxies, the average density is roughly 7×10^{-31} gm/cm^3. But there are reasons to believe that other unseen matter may be present, and that the value of the density may be approximately 2×10^{-29} gm/cm^3 consistent with $v_0^2 = 0$. In this case eq. (4) is easily integrated to give for the zero pressure case

$$1 = 6\pi G \rho t^2, \tag{7}$$

and $r \sim t^{2/3}$. For this zero pressure case the expansion parameter, eq. (3), is equal to 0.5.

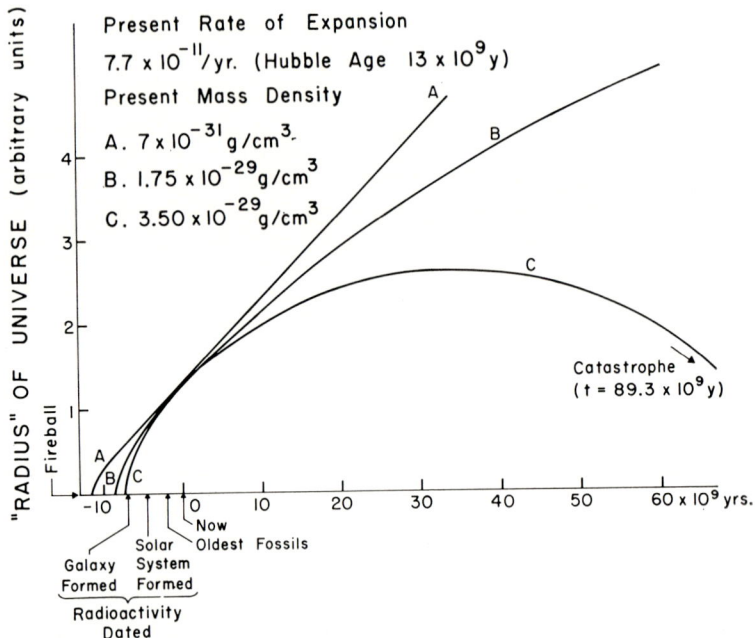

FIG. 27. Cosmology under unmodified General Relativity. The "radius" of the Universe is a measure of the distance between nearest galaxies. The Hubble age would be the age of the Universe if it expanded uniformly (without deceleration).

Figure 27 shows curves representing solutions for r as a function of t with three different choices of present mass density (and $P = 0$), but the same present value of the Hubble age.

As the discussion below will support, there are reasons for believing that the pressure of matter in the early Universe was not negligibly small, that on the contrary it satisfied the relation $P = \frac{1}{3} \rho c^2$. This is the maximum pressure possible for ordinary matter and is characteristic of relativistically moving particles such as electromagnetic radiation. For this case, eq. (5) gives $\rho\, r^4 \sim$ constant and eq. (7) becomes

$$1 = \frac{32\pi}{3} G\rho t^2 \qquad (8)$$

$$r \sim t^{1/2}.$$

Under the scalar-tensor theory of gravitation the scalar field is even more incompressible than electromagnetism or high-temperature matter. If the energy content of the Universe is predominantly that of the scalar field, the pressure is $P = \rho c^2$, the highest pressure possible under relativity. In this case eq. (7) becomes

$$1 = 24\pi G\rho t^2$$
$$r \sim t^{1/3}. \tag{9}$$

It is believed that we see only part of the Universe when we look out into space. In principle, light rays traveling unscattered from the farthest reaches of the "seen Universe" would show the Universe at a very early age. In practice, this view of the very young Universe is not easily obtained. Distant galaxies are both very dim and red. Even the extremely bright quasars are not observed with red shifts much greater than a factor of 3 increase in wave length.

As the Universe grows older, and light travels farther, the radius of the visible Universe recedes faster than matter at its boundary and more matter comes into view. For a Universe with v_o^2 negative the whole universe is seen finally at the moment of final implosion, but even here the total matter seen is finite, for such Universes are finite being limited by a closed space to a finite volume.

There are peculiar puzzles about this Universe of ours. As it gets ever older, more and more of the Universe comes into view, but when new matter appears it is isotropically distributed about us, and it has the appropriate density and velocity to be part of a uniform Universe. How did this uniformity come about if the first communication of the various parts of the Universe with each other first occurred long after the start of the expansion? C. Misner suggests that the extremely young Universe was highly disorganized and that, because of the disorder, the various parts were brought into contact with each other before the disorder was eradicated.

Another matter is equally puzzling. The constant v_o^2 in eq. (4) is very small, so small that we are uncertain with our poor knowledge of ρ as to whether or not it is zero. But the first term on the right of eq. (4) was very much larger much earlier, at least 10^3 times as great when the galaxies first started to form and at least 10^{13} times as great when nuclear reactions were taking place in the "fireball," assuming that the "fireball" story to which I shall return is correct. The puzzle here is the following: how did the initial explosion become started with such precision, the outward radial motion became so finely adjusted as to enable the various parts of the Universe to fly apart while continuously slowing in the rate of expansion?

There seems to be no fundamental theoretical reason for such a fine balance. If the fireball had expanded only .1 per cent faster, the present rate of expansion would have been 3×10^3 times as great. Had the initial expansion rate been .1 per cent less and the Universe would have expanded to only 3×10^{-6} of its present radius before collapsing. At this maximum radius the density of ordinary matter would have been 10^{-12} gm/cm^3, over 10^{16} times as great as the present mass density. No stars could have formed in such a Universe, for it would not have existed long enough to form stars.

No attempt will be made here to discuss quantitatively in an adequate way the complexities introduced into cosmology by the scalar-tensor theory. Perhaps the most interesting aspect of the scalar-tensor cosmology is the gradual increase with time of the scalar as it is generated by the matter content of the Universe[8] (see figure 28). For a cold Universe (zero pressure), the rate of increase of the scalar with time is

$$\frac{d\phi}{dt} = \frac{8\pi}{2\omega + 3}\rho t. \qquad (10)$$

where ω is the dimensionless constant (~ 5) discussed

---

[8] C. Brans and R. H. Dicke, *Phys. Rev.* **124** (1961): p. 925.

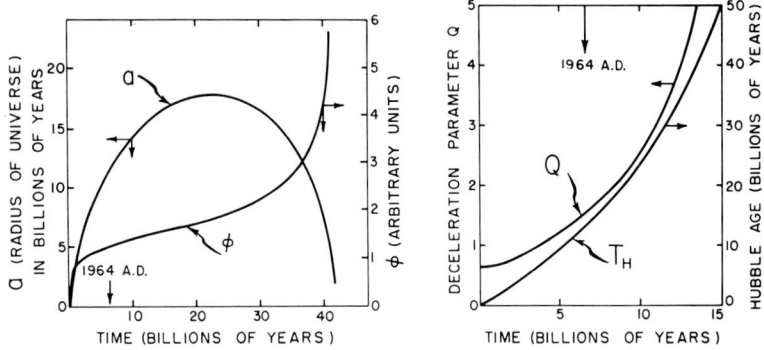

FIG. 28. Similar to curve C of figure 27, except that the gravitational constant is described by the scalar-tensor theory, a general relativistic theory similar to Einstein's theory. ϕ is the scalar which determines the strength of gravitation, being proportional to the reciprocal of the gravitational constant.

earlier. The greater the mass density and the older the Universe, the more rapid is the rate of growth of ϕ. From eq. (10)

$$\phi = \frac{8\pi}{2\omega + 3} <\rho t> t. \qquad (11)$$

where the bracket means a time average of ρt. In this theory the locally measured value of the "gravitational constant" is not constant but is determined by ϕ, being equal to

$$G = (2\omega + 4)/\phi(2\omega + 3). \qquad (12)$$

It should be noted that the greater the density and age of the Universe the smaller the resulting gravitational constant. In this theory the gravitational force is weak in comparison with the electrical force because the visible universe contains so much matter. The ratio of the electrical to gravitational interaction between two electrons is

$$\frac{e^2}{Gm^2} = \frac{4\pi}{\omega + 2} \cdot \frac{c^2 <\rho t> tr_o}{m} \sim \frac{M}{m} \cdot \frac{r_o}{R} \qquad (13)$$

where $r_o = e^2/mc^2$ is the classical radius of the electron, and M and R are the mass and radius of the "visible Universe."

Under the scalar-tensor theory assuming that $\rho \sim 2 \times 10^{-29}$ gm/cm^3, the critical mass density for which $\ddot{r}_o = 0$, r and ϕ vary with time as

$$r \sim t^{(2\omega+2)/(3\omega+4)}$$
$$\phi \sim t^{2/(3\omega+4)} \tag{14}$$

Assuming that $\omega \sim 5$, eq. (14) implies a gradual weakening of gravitation, $\sim 10^{-11}$ parts per year.

The influence of the past history of the seen part of the Universe in determining the present value of the gravitational constant is of significance for Mach's principle.[9] This somewhat imprecise idea of Mach's attempts to associate inertial effects observed locally with distant matter. Under the scalar-tensor theory, the greater the matter content of the Universe and the closer this matter lies to the test body the greater the inertial forces are in relation to gravitational forces, i.e. the weaker is gravitation.

In considering the early expansion of the Universe under the scalar-tensor theory, there are two different effects which should be noted. First the larger value of the gravitational constant earlier increases the deceleration of the expansion. Secondly the possible presence of a large scalar contribution to the energy density affects the expansion rate (eq. (4)).[10] I shall return to these matters later.

The Cosmic Fireball

The idea that the Universe might have had a hot origin, starting out as a radiation-dominated space is quite old. The basic cosmological theory was investigated as early as 1934,[11] but Gamow and his student Alpher were the first

---

[9] E. Mach, *Science of Mechanics* (1883; republished by Open Court Publ. Co., Chicago, 1902), see chapt. 2, sec. 6.
[10] R. H. Dicke. *Astrophys. Jour.* **152** (1968) : p. 1.
[11] R. C. Tolman, *Relativity, Thermodynamics and Cosmology* (Oxford University Press. 1934).

seriously to consider the various physical implications of a Universe with a hot beginning.[12] This work was motivated by Gamow's idea that the heavy element might have arisen through nuclear reactions in the young hot Universe. This led to a series of developments, but ultimately it was found that significant amounts of heavy elements could not be formed in this way.[13] The hot cosmology thereafter fell into disfavor and almost disappeared from view, but it was later revitalized by Hoyle and Taylor to provide a means for the formation of helium.[14]

Without having seen the Hoyle and Taylor paper and with a remarkable ignorance of the development that had been initiated by Gamow, several of us at Princeton independently suggested a hot origin of the Universe.[15] But the motivation was quite different. We were impressed by the arguments in support of an oscillating Universe, but such a Universe seemed to require a hot "bounce." To eliminate from the Population II stars the heavy elements formed by stars during the previous expansion of the Universe, temperatures in excess of $10^{9°}$ K seemed to be required thermally to decompose the nuclei into elementary particles. Furthermore, the gravitational collapse of our Universe with its present radiation content would lead to the adiabatic compression of the radiation and the formation of an extremely hot fireball.

Such a hot early phase of the Universe, the cosmic fireball phase, implies the present existence of residual thermal, or black-body radiation, and P. Roll and D. Wilkinson of our research group started to build an apparatus to look for the radiation. Before the apparatus could be tried we had

---

[12] G. Gamow, *Phys. Rev.* **70** (1946) : p. 572; R. A. Alpher, *Phys. Rev.* **74** (1948) : p. 1577.

[13] G. Gamow. *Phys. Rev.* **74** (1956) : p. 505; R. A. Alpher, H. A. Bethe, and G. Gamow, *Phys. Rev.* **73** (1948) : p. 803; R. A. Alpher, J. W. Follin, and R. C. Herman, *Phys. Rev.* **92** (1953) : p. 1347; R. A. Alpher and R. C. Herman, *Rev. Mod. Phys.* **22** (1950) : p. 153; E. M. Burbidge, G. R. Burbidge, W. A. Fowler, and F. Hoyle. *Rev. Mod. Phys.* **29** (1957) : p. 547.

[14] F. Hoyle and R. J. Tayler, *Nature* **203** (1964) : p. 1108.

[15] R. H. Dicke, P. J. E. Peebles, P. G. Roll, and D. T. Wilkinson. *Astrophys. Jour.* **142** (1965) : p. 414.

a call from A. A. Penzias who told us about the unexplained receiver noise that he and R. W. Wilson had been observing.[16] We and they now believe that this noise had its origin in thermal radiation entering the antenna.[15] The existence of cosmic thermal radiation had been earlier suggested by Gamow, and Alpher and Herman[19] had made a reasonably good estimate of the present temperature.

After Roll and Wilkinson made their first measurements[17] which supported the interpretation of the radiation as thermal, a long series of observations was made by various groups. These have given a present temperature of 2.7° K for the radiation.[18] Recently a rocket-borne apparatus has given a radiation flux at sub-millimeter wave lengths which may not be consistent with the thermal spectrum.[18a] This radiation may be of local origin in the Galaxy.

It is ironical that, owing to the theoretical difficulties, the original motivation for our work on cosmic black-body radiation is now somewhat in question. It has long been known that the cosmological equations cannot be integrated through the collapse of the Universe to show the existence of a "bounce" leading to the start of a new expansion. A mathematical singularity develops in the solution. It had been thought that this singularity was due to the over-idealized nature of the physical assumptions of isotropy and uniformity. It was hoped that the singularity would disappear if sufficient irregularity were introduced into the mathematical model.[20] This has not occurred and the

---

[16] A. A. Penzias and R. W. Wilson, *Astrophys. Jour.* **142** (1965): p. 419.

[17] P. G. Roll and D. T. Wilkinson, *Phys. Rev. Ltrs.* **16** (1966): p. 405; *Ann. of Phys.* **44** (1967): p. 289.

[18] T. F. Howell and J. R. Shakeshaft, *Nature* **216** (1967): p. 753; W. J. Welch, et al., *Phys. Rev. Ltrs.* **18** (1967): p. 1068; M. S. Ewing. B. F. Burke, and D. H. Staelin, *Phys. Rev. Ltrs.* **19** (1967): p. 1251; D. T. Wilkinson. *Phys. Rev. Ltrs.* **19** (1967): p. 1195; R. A. Stokes, R. B. Partridge, and D. T. Wilkinson, *Phys. Rev. Ltrs.* **19** (1967): p. 1199; P. E. Boynton, R. A. Stokes, and D. T. Wilkinson, *Phys. Rev. Ltrs.* **21** (1968): p. 462.

[18a] K. Shivanadan, J. K. Houck, and M. O. Harwit, *Phys. Rev. Ltrs.* **21** (1968): p. 1460.

[19] R. A. Alpher and R. C. Herman, *Nature* **162** (1948): p. 774.

[20] E. M. Lifshitz and I. M. Khalatnikov, *Adv. in Phys.* **12** (1963): p. 183.

mathematical singularity seems to be necessary under present theory.[21]

The difficulty may lie in the theory or it may be that the span of physical time has finite limits. We cannot be sure, but some of us feel that singularities are more the business of mathematicians than physicists, that the singularity appears in our mathematical model because of some inadequacy of the theory.

One strange aspect of the involvement of the Princeton group in the detection and study of the cosmic black-body radiation has not been previously discussed. The apparatus built by Roll and Wilkinson was derived from the microwave radiometer which I had developed during World War II.[22] (See figure 29.) This old radiometer had been used to measure the absorption of microwaves by atmospheric water vapor.[23] But incidental to these measure-

Fig. 29. The first microwave radiometer, built in early 1945. An upper limit of 20° K for the temperature of cosmic thermal radiation was obtained in 1945, also isotropy of the radiation to 1° K.

---

[21] S. W. Hawking and G. F. R. Ellis, *Astrophys. Jour.* **152** (1968) : p. 25.
[22] R. H. Dicke, *Rev. Sci. Instr.* **17** (1946) : p. 268.
[23] R. H. Dicke, R. Beringer, R. L. Kyle, A. B. Vane. *Phys. Rev.* **70** (1946) : p. 340.

ments I had obtained an upper limit of 20° K for the temperature of "radiation from cosmic matter" at 1.25 cm wave length.[23] Although I had personally made the analysis that gave this result, this had been completely forgotten.

The early result was found by accident by Peebles in the original publication. But after being reminded of it I also remembered that an upper limit of 1° K to the anisotropy of radiation of this type had also been measured in 1945. (This result had not been published.)

I must emphasize that in 1945 when these measurements were made we did not have a hot cosmology in mind, but rather the accumulated radiation from distant galaxies.

Gamow's 1946 paper in the *Physical Review*[12] appeared only 232 pages after ours but there was no communication in either direction. While our 20° K limit was crude, even this was a useful result for it showed that the energy density of thermal radiation is not great enough to stop the expansion of the Universe.

FIG. 30. The original Roll-Wilkinson radiometer housed on top of the Princeton biology building under an old "bird platform."

Fig. 31. The three Princeton precision radiometers which operate at the wave lengths 3.2, 1.58, and 0.856 cms. They are located at an altitude of 12,470 ft. at the Barcroft facility of the White Mountain Research Station, Bishop, California.

In figures 30 and 31 are shown the original Roll-Wilkinson radiometer and recent Princeton precision radiometers. The three most accurate radiometer measurements of the cosmic radiation are indicated on figure 32 along with the others.[24]

A matter of considerable importance for cosmology is the isotropy of the thermal radiation. This has been studied extensively by Partridge and Wilkinson.[25] (See figure 33.) Their results are 2 orders of magnitude more precise than

---
[24] D. T. Wilkinson, *Phys. Rev. Ltrs.* **19** (1967): p. 1195; R. A. Stokes, R. B. Partridge, and D. T. Wilkinson, *Phys. Rev. Ltrs.* **19** (1967): p. 1199.

[25] R. B. Partridge and D. T. Wilkinson, *Phys. Rev. Ltrs.* **18** (1967): p. 577; *Nature* **215** (1967): p. 719.

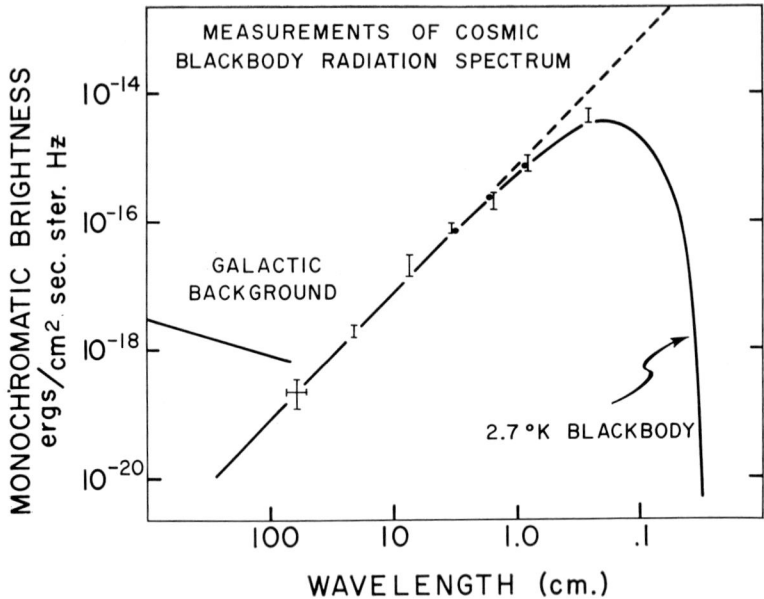

FIG. 32. Spectral distribution expected for thermal (black-body) radiation at a temperature of 2.7° K. The three precision measurements are shown as dots. Their errors are too small to show on the figure.

my 1945 measurement. Their measurements indicate that the cosmic thermal radiation is isotropic to better than 10^{-2}° K.[25] An upper limit of about 300 km/sec can be set for the velocity of the earth relative to the radiation by these isotropy measurements.[25]

The general theory of the cosmic fireball indicates that the temperature of the radiation falls with time, varying inversely as the radius of the Universe. This seems paradoxical to many physicists, for this implies that the number of photons in the Universe is constant but that the energy of each photon decreases with time, varying inversely as the radius of the Universe. It seems paradoxical that the freely propagating, non-interacting photons should lose energy. The reason this energy loss occurs can be seen in two ways:

1) The photons we observe now were emitted long ago in the young Universe by matter moving rapidly away from

us. This rapid motion causes a Doppler effect, a shift to the red. The red shift is greater the older the Universe, for the older the Universe the farther away from us is the point of origin of the photons. But the velocity of recession of the matter at the point of origin is greater the greater the distance.

2) The Universe is uniform, hence symmetric, an imaginary plane cutting through the Universe divides it into two equivalent parts. Equal fluxes of thermal radiation fall on both sides of the surface. The surface acts somewhat like a perfectly reflecting mirror, in the limited sense that the incident flux and the "reflected flux" are equal. Construct an imaginary box of such imaginary "perfectly reflecting" walls. The box expands in volume with the expansion of the Universe and the internal radiation energy decreases because of "work" done by radiation pressure on

FIG. 33. The Princeton isotropy radiometer, constructed by Partridge and Wilkinson. This was operated for a year at one of the world's most unattractive sites, the desert at Yuma, Arizona.

the "walls." Alternatively, standing waves in such a box would be stretched along with the box, and wave length and the radius of the Universe should vary together.

It is possible to picture in bold outlines the metamorphosis of the cosmic fireball into the Universe as we know it today.[15] While this picture is based on sound theoretical analysis, it must be emphasized that the observational basis for the analysis is meager. The picture could be completely wrong.

We start our narrative when the fireball has a temperature of 10^{10}° K. Here all nuclei are thermally decomposed into elementary particles, and protons and neutrons are present in nearly equal numbers. Electrons, positrons, and photons are also present, roughly 10^8 times as abundant as protons and neutrons.

As the Universe expands the temperature of the fireball falls. When the radius has increased by a factor of 10 (in 180 seconds) the temperature falls to 10^9° K where neutrons and protons can combine to form deuterons without being immediately decomposed thermally. Once deuterons are formed other nuclear reactions can take place to form helium.[26] We shall return to this question of possible helium formation.

Quite by coincidence another important effect occurs at temperatures of about 10^9° K. The enormous numbers of positrons are annihilated by electrons to generate photons. This increases the number of photons by approximately a factor of two.

The Universe, now containing ionized hydrogen and possibly helium, continues to expand. But after expanding by an additional factor of 2.5×10^5, the temperature having fallen to 4×10^3° K (1×10^5 years after the start of the expansion), the electrons and nuclei of the ionized plasma combine to form neutral gas. At this point a new phe-

---

[26] P. J. E. Peebles, *Astrophys. Jour.* **147** (1967) : p. 859; R. V. Wagoner, W. A. Fowler, and F. Hoyle, *Astrophys. Jour.* **148** (1967) : p. 3.

nomenon takes place. Small irregularities in gas density in the space start to increase rapidly.[27] This leads to the fragmentation of the gas into separated clouds. These clouds are expected to be of substantially the same size with a limited range of mass. The characteristic mass of the clouds depends upon the poorly known present mass density of the Universe, and also upon gravitational theory, but this mass is independent of the initial density irregularities providing they are randomly positioned and of very small scale, a characteristic volume element containing less than 10^4 solar masses. (See figure 34.)

As the Universe continues to expand these clouds stay constant in size but move apart with the expansion of the Universe. Ultimately the randomness in their positions

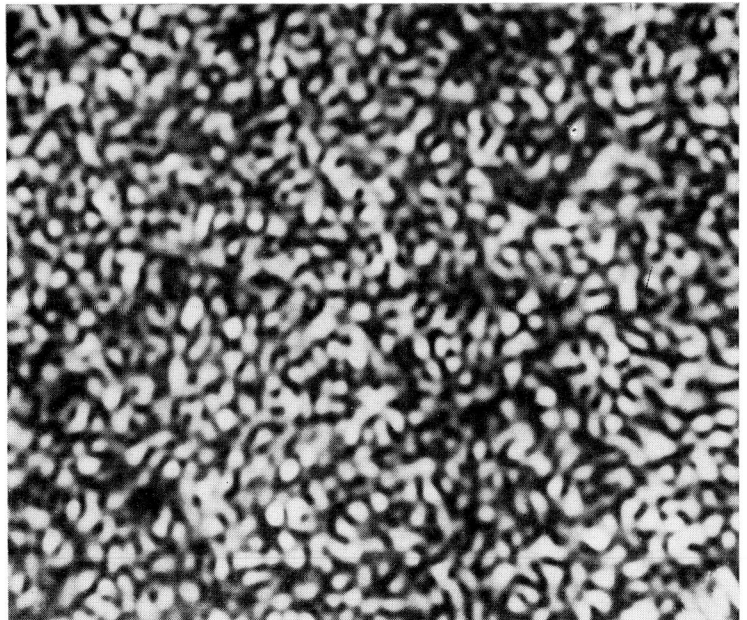

Fig. 34. A computed pattern representing the density distribution in the cooling fireball as gas clouds begin to form. We believe that the globular clusters represent clouds of this type fragmented into stars.

[27] G. Gamow. *Phys. Rev.* **74** (1948) : p. 505; P. J. E. Peebles, *Astrophys. Jour.* **142** (1965) : p. 1317; *ibid.* **147** (1967) : p. 859.

causes groups of 10^2 to 10^6 of the clouds to become gravitationally bound, to stop expanding, and to start contracting some 200 million years after the fireball phase.

Before a cataclysmic collison of these clouds with each other a few manage to convert their gas into stars. These are believed to represent the globular clusters.[28] Collisions between the remaining clouds initiate star formation and most of the remaining gas is believed to be converted to the Population II stars at this time. These stars are believed to be still moving with the distribution of velocities present in the gas clouds at the time the stars were formed.

The original gas clouds should have possessed little spin angular momentum for the hot, ionized fireball had a high viscosity and any postulated primeval turbulence in the fireball would have been quickly damped.[28]

The observed properties of globular clusters are remarkably similar to the computed properties of the gas clouds. The clusters are remarkably spherical, indicating little spin angular momentum. Their luminosities and probably their masses vary by only about a factor of three. These luminosities are substantially independent of the positions of the clusters and seem to be unchanged even when the clusters are found on other galaxies. The shapes and diameters of globular clusters vary little.

To recapitulate briefly, we believe that the Universe as we know it developed from a very hot fireball which cooled as it expanded. At a temperature of $10^{9°}$ K helium may have been formed. At a temperature of $4 \times 10^{3°}$ K the electrons and nuclei in the ionized plasma combined to form neutral gas. Gas clouds condensed in this neutral gas. Gravitationally bound groups of these clouds ultimately formed galaxies. We believe that globular clusters represent the fossil remains of the original gas clouds.

---

[28] P. J. E. Peebles and R. H. Dicke, *Astrophys. Jour.* **154** (1968) : p. 891.

The Cosmic Fireball and the Scalar-Tensor Theory

Our knowledge of the history of the cosmic fireball is meager, and no secure test of gravitational theory is presently possible using this knowledge. This inaccuracy of the observations is partially compensated for by the magnitude of the relativistic effects expected under the scalar-tensor theory during the fireball phase, but the reliability of the interpretation of some of the observations is another matter. While no firm conclusions can be presently drawn, in order to help interpret future observations we shall examine the implications of the scalar-tensor theory of gravitation for the cosmic fireball.

It has previously been noted that the influence of the scalar field on the development of the Universe is presently minimal. The fractional rate of decrease of the "gravitational constant" associated with the scalar field is of the order of 10^{-11} per year, too small to induce significant effects.

The theory of the cosmic fireball interpreted under the scalar-tensor theory of gravitation has been developed.[29] Earlier, the gravitational constant would have been at least a factor of 5 greater for a closed space (a present mass density of 2×10^{-29} gm/cm^3). If the dominant energy content of the Universe at this early time were that of the scalar field, the "gravitational constant" could be in excess of 10^5 times the present value with a scalar energy density in excess of 10^8 times that of the fireball radiation. (See figure 35.)

As was noted before, we are very uncertain of the present mass density. If this is as low as 7×10^{-31} gm/cm^3 the corresponding lower limit on the "gravitational constant" at the time of helium formation would be 2 times the present value, but again it could be much greater.

Two aspects of the development of the fireball are of particular concern here. Assuming the validity of the over-

---

[29] R. H. Dicke, *Astrophys. Jour.* 152 (1968): p. 1; G. S. Greenstein, *Astr. and Space Sci.* 2 (1968): p. 155.

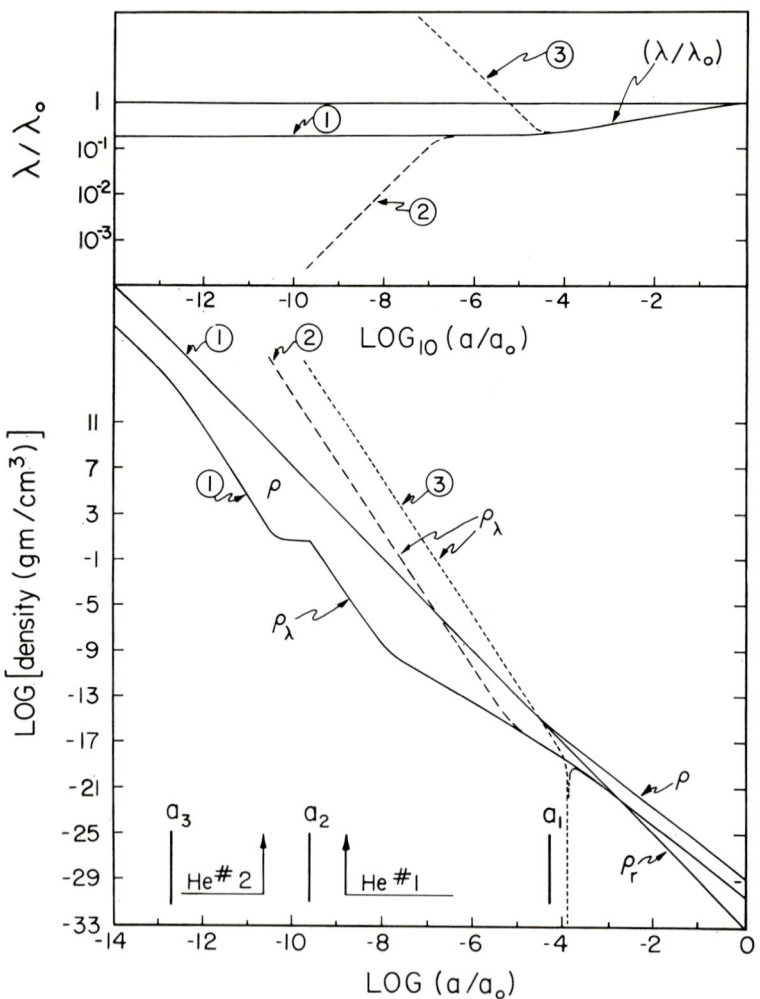

FIG. 35. The evolution of the hot Universe computed under the scalar-tensor theory assuming a Hubble age of 10^{10} years, a present mass desnity of 2×10^{-29} gm/cm^3, and a present radiative temperature of 3° K. ρ refers to the sum of matter and radiation density, expressed in gms/cm^3. ρ_r and ρ_λ refer respectively to the radiation field and the scalar field. Solution 1 is that for which the Universe starts expanding with the minimum scalar field energy possible. For solution 2 the early Universe is dominated by scalar field energy. Solution 3 is non-physical.

all picture, both the formation of helium[29] and the formation of globular clusters[28] would be expected to be significantly influenced by the type of gravitational theory employed.

Consider first the formation of helium.[26] When the temperature was as high as $10^{10°}$ K, protons and neutrons were equally abundant. The normally expected radioactive decay of neutrons into protons and electrons did not occur because equal numbers of inverse decays occurred through the collision of energetic electrons of the hot plasma with protons. As the plasma temperature fell to $10^{9°}$ K, the thermal equilibrium shifted strongly toward the proton state, but the temperature drop was so rapid that equilibrium was not reached. There was insufficient time for the neutrons to decay to protons, and enough neutrons may have remained to permit the formation of some helium. Assuming the validity of General Relativity and the suitability of the cosmological mode, 25-28 per cent of the gas would have formed helium.

The expected yield of helium depends sensitively upon the rate of change of the temperature, hence the expansion rate of the Universe.[30] If the rate of expansion is increased, there is less time for neutrons to decay and more neutrons are available to induce the formation of helium. But too rapid an expansion rate does not provide sufficient time for nuclear reactions to occur. Figure 36 is taken from Peebles' paper.[30] It shows the effect of expansion rate on the yield of helium. By increasing the expansion rate by a factor of 100 the helium yield rises to over 80 per cent, but at a rate 10^5 times as great, the yield is substantially zero.

The expansion rate varies as $\sqrt{G\rho}$ (see eq. 4 remembering that v_o^2 is negligible). Under the scalar-tensor theory of gravitation, assuming a present mass density of 2×10^{-29} gm/cm^3, the expansion rate would be at least 2.3 times as great as under General Relativity but the rate could be over

---
[30] P. J. E. Peebles, *Astrophys. Jour.* **146** (1966) : p. 542.

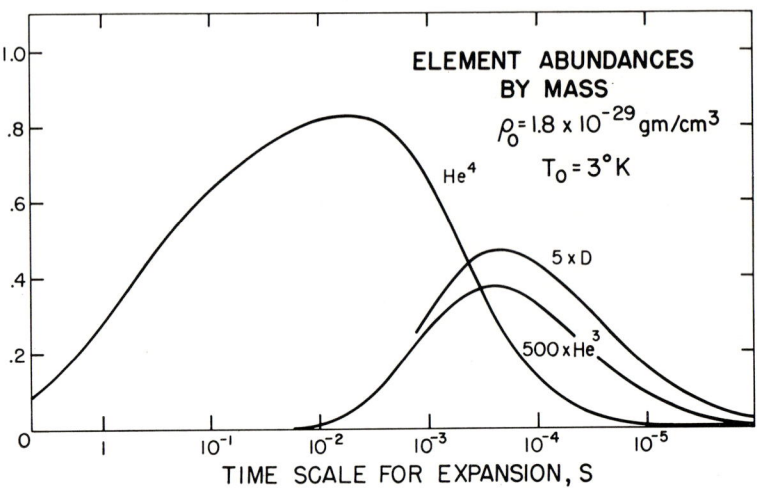

FIG. 36. The effect of expansion rate on the production of helium, deuterium, and "helium three" in the cosmic fireball (from Peebles). $S = 1$ corresponds to General Relativity. Smaller values of "S" represent more rapid expansion. Under the scalar-tensor theory, the same present density $(2 \times 10^{-29} \text{ gm/cm}^3)$, radiation temperature 3° K, and assuming a Hubble age of 10^{10} years, $S < 0.45$. This is indicated on the graph.

3×10^6 times as great. The expected yield of helium in the first case is 42 per cent and it is 0 per cent for the case of extremely rapid expansion.[29] Assuming a present mass density of 7×10^{-31} gm/cm^3, the corresponding helium yields would be 32 per cent and 0 per cent respectively.[29]

There are no known naturally occurring mechanisms in the Galaxy for decomposing large amounts of helium to hydrogen, whereas stars are known to burn hydrogen to helium, the chief source of their heat. We expect to find a ratio of helium to hydrogen as great or greater than that obtained from the fireball. The sun seems presently to have less than 30 per cent helium at its surface and the 42 per cent derived above is too large. It may be that the only possibility compatible with the observations for the high density Universe is for no helium to be produced in the fireball. While figure 36 seems to indicate that 20-30 per cent yield of helium is possible for $S \sim 3 \times 10^{-4}$, this can be excluded on the grounds that the large yield of deuterium expected under this condition is not observed.

The presence of a scalar component in gravitation is not the only way that the yield of helium from the fireball can be affected. Lack of uniformity and isotropy of the fireball tends to increase the expansion rate, but an increase by the factor of 10^5 needed to stop helium formation does not seem possible. An enormous unbalance of neutrinos relative to antineutrinos could also affect helium production,[15, 26] but the required excess is so great, $\sim 10^7$ times the proton abundance, that it seems unlikely.

Spectroscopic observations show that, by mass, approximately one-fourth of the interstellar gas is helium, but it is difficult to decide whether this helium was formed in the hot fireball or much later, along with the other elements found in Population II stars.

The outer layers of a Population II star is an excellent place to look for primeval helium, but Population II stars on the main sequence are too cool to excite helium lines. While there are a few hot Population II stars, these are all highly evolved and their outer layers may have been contaminated by helium from the burned out core.

It is interesting that these hot blue stars show very weak helium lines, but the significance of the lines is unclear, for weak lines are also found occasionally in blue Population I stars known to contain substantial amounts of helium.

Another way in which the helium content of Population II stars might be determined is via the connection between the mass, mean molecular weight, and luminosity of a star. The calculated luminosity is sensitive to the average molecular weight, varying approximately as the seventh power of the molecular weight. If both the mass and luminosities are known, the mean molecular weight and the helium content could be determined. Unfortunately, of the dozen or so stars whose masses are reliably known, none are Population II.

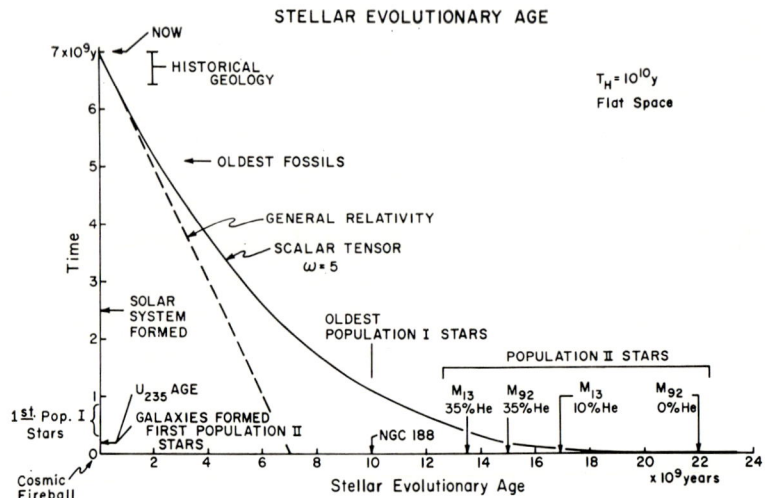

FIG. 37. Stellar evolutionary ages computed assuming the validity of General Relativity are plotted on the horizontal axis. The solid curve gives the relation with age under the scalar-tensor theory (or equivalently as shown, the time of formation).

The matter of helium content is of considerable importance, for the computed age of these oldest stars varies by almost 50 per cent depending upon the assumed helium content. See figure 37, but please note that recently published ages have tended to be a few billion years less than those shown. The stellar evolutionary ages computed and plotted along the horizontal axis of figure 37 are ages based on the assumption that gravitation is constant. Under the scalar-tensor theory the luminosity is greater in the past, for the luminosity varies approximately as the gravitational "constant" to the seventh power. Thus the computed ages are less under this theory.[31] The solid curve of figure 37 provides a connection between the evolutionary age computed under General Relativity and the time of origin computed under the scalar-tensor theory, assuming a Hubble age of the Universe of 10^{10} years and a present mass density of 2×10^{-29} gm/cm^3.

---

[31] R. H. Dicke, *Rev. Mod. Phys.* 34 (1962): p. 110, chapt. of *Stellar Evolution*, R. F. Stein and A. G. W. Cameron, eds. (N. Y., Plenum Press, 1966).

It should be noted that under the scalar-tensor theory an evolutionary age of 15×10^9 years would be appropriate for Population II stars (or 2×10^{10} if the Hubble age is 13×10^9). An evolutionary age as great as 2×10^{10} probably requires very low helium, consistent with expectations under the scalar-tensor theory. However, under General Relativity the expected helium content is 27 per cent and the age is less, perhaps as low as 10^{10} which is consistent with a Hubble age of 15×10^9. Both interpretations are reasonably consistent with the present expansion rate of the Universe[5] (which determines the Hubble age) and the uranium decay age of the Galaxy.[4]

When the temperature of the fireball has fallen to 4000° K, at which the plasma becomes neutralized and gas clouds begin to form, the gravitational "constant" under the scalar-tensor theory is 3.4 times its present value, for a present mass density of 2×10^{-29} gm/cm³ and a Hubble age of 10^{10} y (with $\omega = 5$). For a present density of 7×10^{-31} gm/cm³ the gravitational "constant" would have been 1.75 times as great. The mass of clouds formed from the gas depends upon the gravitational constant at that time, and also upon the average density of the Universe.

The average size of these proto-cluster clouds is given approximately by the Jeans length[27, 28]

$$\lambda = (\pi k T / G \rho m)^{1/2} \tag{15}$$

where m is the mass of the hydrogen atom, $T = 4000°$ K is the temperature of the gas, and ρ is the gas density at that temperature. Their average mass is approximately

$$\rho \lambda^3 = \rho^{-1/2} \left(\frac{\pi k T}{G m} \right)^{3/2}. \tag{16}$$

For the two assumed present mass densities 2×10^{-29} gm/cm³ and 7×10^{-31} gm/cm³ the globular cluster masses expected under General Relativity are respectively 6×10^5 and 3×10^6 solar masses. Under the scalar-tensor theory the corresponding masses are 1×10^5 and 1.5×10^6 solar masses respectively.

With the assumptions that the origin of globular clusters has been correctly described and that no appreciable mass has been lost from the clusters, the observed masses of clusters might help us determine the present density of the Universe and hopefully the correct gravitational theory.

But this is being too optimistic. Many astronomers are doubtful about this theory of globular clusters, and there is considerable doubt about stellar evaporation from globular clusters. Furthermore, there are only two globular clusters whose masses are known. These are of approximately $(1.4 \text{ and } 2.5) \times 10^5$ solar masses. While these masses tend to favor the larger density for the Universe and the scalar-tensor theory, this interpretation is highly uncertain.

In conclusion, the implications of the scalar-tensor theory are most significant for the young Universe. We have noted that the expected production of helium in the fireball depends significantly upon gravitational theory. The expected average mass of globular clusters is significantly affected by this change in theory. The effect most likely to show up first in our observations, if the scalar-tensor theory is valid and the present mass density is as great as 2×10^{-29} gm/cm^3, is the low helium in Population II stars. But even this result would not be conclusive in its implication.

If there were other reasons for accepting the scalar-tensor theory, it would represent an extremely useful key to the interpretation of the observations of the cosmic ball, but I doubt that the observations of the cosmic fireball will ever supply the essential element needed for the confirmation or rejection of the scalar-tensor theory.

QC178 .D48
Dicke / Gravitation and the universe